The Social Biology of Ants

Klaus Dumpert
J-W Goethe Universität

Translated by
C Johnson
Oxford University

π

Pitman Advanced Publishing Program
BOSTON · LONDON · MELBOURNE

PITMAN PUBLISHING LIMITED
39 Parker Street, London WC2B 5PB

PITMAN PUBLISHING INC
1020 Plain Street, Marshfield, Massachusetts

Associated Companies
Pitman Publishing Pty Ltd., Melbourne
Pitman Publishing New Zealand Ltd., Wellington
Copp Clark Pitman, Toronto

Library of Congress Cataloging in Publication Data
Dumpert, K
    The social biology of ants.
    (The Pitman international series in neurobiology & behaviour)
    Translation of Das Sozialleben der Ameisen.
    Bibliography: p.
    Includes index.
    1. Ants-Behaviour. 2. Social behaviour in animals.
    I. Title. II. Series: Pitman international series in neurobiology & behaviour.
    QL568.F7D8513   595.7'96   79-25013

Translated by C. Johnson from DAS SOZIALLEBEN DER AMEISEN

© Verlag Paul Parey 1978

First German edition published by Verlag Paul Parey, Berlin and Hamburg
First English edition published by Pitman

This translation ©,Pitman Publishing Ltd 1981

Printed and bound in Great Britain
at The Pitman Press, Bath

ISBN 0 273 08479 8

Contents

Foreword

All existing species of ants belong without exception to the societal insects. Ants have been particularly successful with their many species, their worldwide distribution and their corresponding manifold adaptations to different environments. The secret of this success is very closely linked with their social life, being their special fascination. The lively interest which has been directed towards ants up to the present day is reflected in the unusually large number of works about them.

In contrast to the abundance of these individual investigations, however, the number of summarial writings is very limited, especially in the German-speaking world. There is, in fact, no recent comprehensive work dealing with all the different aspects of the social life of ants. The author seeks to bridge this gap with the following book, which is addressed not only to biologists in various fields including biology lecturers and students, but also to all who have an interest in social insects, in particular, the social life of ants. Given the large number of works which have appeared on the subject, however, it has become essential to limit the use of material and arrive at a selection. The author has, therefore, limited the different aspects of social life in his presentation, whilst other aspects, such as the ecological role of ants and their significance in issues of hygiene and economics, which have only an indirect bearing upon the subject in hand, have not been discussed.

In his presentation of the social life of ants, the author has been obliged to impose some restrictions necessitated by the limited scope of this publication. Essentially, therefore, importance was attached to limiting the scope of the theme and drawing attention to the different aspects. Readers interested in a more exact account of these are referred to the sources in the bibliography.

The author has been generously supported in the work on this book by a number of people, and would like to take this opportunity to thank them. He is particularly grateful to Professor U. Maschwitz, Professor A. Buschinger and Professor St Vogel, who read either parts, or the whole of the manuscript, made valuable suggestions and provided photographic illustrations. I would also like to thank Mrs C. Nitsche for her most careful completion of the drawings, Professor Gnatzy for his preparation of photographs from the scanning electron microscrope, and Paul Parey Verlag — especially C. Georgi — for their sympathetic co-operation. Last but not least, I would like to thank my wife for her help with proofreading.

Königstein/Ts., March 1977   *Dr K. Dumpert*

1 Introduction

1.1 Special features of ants

'Go to the ant, thou sluggard; consider her ways, and be wise.' (Proverbs, Ch. 6,6-8) This rather apt reference to ants is in the Old Testament. Readers of this text in obvious need of moral improvement are called upon to take instruction from the behaviour of ants. Above all, the text reveals what it was about ants that drew the attention of men of earlier times. The observer is struck more than anything else by the industriousness with which these creatures harvest corn in the summer, as is shown later in the text. The industry of the ant is still a byword today: in German the word for industry is derived from the word for ant; the Latin word, 'formica', recalls the activity of harvester ants, from 'ferre', meaning to carry, and 'mica', corn.

Industriousness as a typical characteristic of ants should not, however, stand as a central idea for this book, since there are also 'lazy' ants, i.e. workers who specialize in doing nothing. The single typical characteristic of all ants in the Formicidae family is their social life. Apart from the colony founding females of certain species, there are in fact no solitary ants.

By social insects in the true sense of eusociality, one is describing a situation in which several generations live together and in which the offspring are reared by a caste itself unable to reproduce, so that one is able to distinguish a division of labour. These criteria of eusociality are only fulfilled by relatively few insect groups, nearly all of which belong to the genus Hymenoptera. Eusocial structures have developed eleven times, independently of one another within the Hymenoptera, two or more times among wasps (Stenogastrinae, Vespinae and probably in addition, in the genus *Microstigmus,* which belongs among the sand wasps, or Sphecidae) at least eight times among bees (e.g. among stingless bees, Meliponae, honey bees, bumble bees and furrow bees), and at least once among ants. Among all other insect groups put together, eusociality evolved in only a single group, the termites (Wilson, 1971).

1

Among the widely differing degrees to which social behaviour manifested in other groups, the habits of the burying beetle come quite close to eusociality. Males and females of this genus *(Necrophorus)* are attracted to carrion and both sexes may then begin to burrow it into the ground. The mate for the ensuing period is acquired at this time and the sexes work together from then on. Fertilization ensues only when the carrion is safely buried in an underground chamber. Among some species, the male departs first and then the female, but generally the sexes remain together. The female lays the eggs and feeds the larvae when hatched, either alone or with the male, on pre-digested food. Towards the end of the larval stage, the parents sometimes burrow passages into the walls of the chambers, in which the larvae then pupate. Small carrion beetles of one species are rather interestingly reminiscent of a sterile worker caste. Several beetles of this species aid with the burying of the carrion, and then leave without taking part in reproduction (Milne and Milne, 1976).

If one measures the level of social organization by the degree of 'selflessness' of its members, their degree of cooperation, their division of labour and the level of solidarity within the group, then the social units of insects by far excel all social forms to be found among vertebrates, with the exception of humans. This demonstrates that social development does not run parallel to other aspects of phylogeny and also shows that an increase in intelligence by no means guarantees the development of social behaviour (Wilson, 1975a).

One explanation for the inverse ratio between the level of phylogenic development and the development of social organization is put forward in the 'kinship' theory of Hamilton (1964). According to this theory, the level of social organization depends on the number of genes held in common by those in the group. One finds the greatest number of commonly held genes in those species of the animal kingdom with the highest level of social organization, that is coral and other, similar, colonies. The offspring of these colonies are nonsexually reproduced and all have identical heredity. Vertebrate animals, on the other hand, reproduce sexually and can only have a maximum of 50% of their genes in common, with the exception of identical twins. This could explain why altruism is rare among the vertebrates, apart from that associated with the rearing of young. Among humans, however, one finds a level of social organization which stands out in glaring contrast to the rest of the vertebrates. Moreover, this level of social organization comes about in a different manner from elsewhere in the animal kingdom, in that it is based on those human abilities made possible by an increase in intelligence, namely the ability to ally oneself with one's rivals and the ability to compromise. The prerequisite for this is a highly differentiated language of symbols, which permits the conclusion of agreements which can be maintained over many generations. Such agreements allow the development of a division of labour which greatly exceeds that of social insects. Typical of human beings, and indicative of their relationship *vis à vis* other vertebrates, however, is the fact that this division of labour is coupled with a preference for the individual rather than the group; it is not based on the altruism of a specific caste, as with the social insects (Wilson, 1975a).

Thus, the social insects stand in the exact middle position between vertebrates and colonies of the coral type, inasmuch as their level of social organization is related to their proportions of common genes. The particular characteristics which determine sex in the Hymenoptera follow from the fact that males issue from unfertilized, haploid eggs, while females and workers develop from fertilized, diploid eggs. This type of sex determination means that female siblings are more closely related to each other than their own offspring. Females of the same parents have a common genetic inheritance of 75%, since they all inherit genes from their haploid fathers, whilst inheriting half the genes of their diploid mothers, but have only 50% of the same genes as their female offspring.

The central point of the 'kinship' theory when applied to the social insects is that their social organization is only made possible by the nature of their kinship relations, in that Hymenoptera societies are chiefly based on the relationship between female siblings, while males have almost no share in the life of the colony.

Whilst it will be admitted in connection with the 'kinship' theory that varied relations within a group will act as foundations for varied development of altruism, which, in the case of social Hymenoptera culminates in the development of an altruistic, sterile worker caste, the objection can nevertheless be raised within the framework of another theory that this sterile worker caste may also evolve through oppression and exploitation (Michener and Brothers, 1971; Alexander, 1974). In other words, individual females may have dominated increasingly over other females in the course of their evolution and finally prevented them from becoming fertile. The oppressed females were forced out of their reproductive role in this manner, and became workers, whilst only the dominant females, the 'queens', reproduced.

There is now the empirical possibility of finding out whether the 'kinship' theory or the 'exploitation' hypothesis is correct (Trivers and Hare, 1976). In colonies where only a reproductive queen can lay eggs, she is the only one who cares for the coming generation. If the 'exploitation' hypothesis is correct, therefore, the ratio of males to reproductive females should balance out in the offspring to a ratio of 1:1. If the 'kinship' theory is correct, however, the flow of workers into the colony, who alone care for the rearing of young, should have become more noticeable in the course of evolution. The workers are more closely related to their female than their male siblings, 75% compared with 25% common genes, and should, as a result of this, care better for their female than their male siblings, in which case a definitive ratio between the sexes would have to average about three females to one male. In fact, through calculations made from large numbers of many species of social ants and bees, a number or at least a weight ratio was found to be three females to one male. Among the solitary, non-social bees, the ratio averaged one male to one female. This marks not only the discovery of a strong argument for the validity of the 'kinship' theory for ants and other social Hymenoptera, but also opens the way for testing the theory empirically for the first time.

Thus, among the social Hymenoptera it was ants which separately developed the most richly varied social life. They manifestly took advantage of all possibilities of social organization in the context of their brain capabilities and the special features of their colony system. That which effected the multiplicity of their social life is possibly also comparable with their most important rivals, the termites (Wilson, 1963).

The worldwide distribution of ants is very closely connected with their social life. Their presence reaches, with a maximal number of species, from the tropics to the Arctic (Gregg, 1972), and extends around the whole world. In relation to humans, this distribution of ants is less reminiscent of morphological and physiological adaptation, than for their attainment of a worldwide distribution across nearly all climactic boundaries, through the development of variations in behaviour patterns.

By way of further peculiarities of ants which are reminiscent of their relationship to humans, one should mention the workers' lack of wings and their somewhat long life expectancy (Kutter, 1969). Owing to their lack of wings, ants, unlike the social bees and wasps, are tied to the ground. They do not have the possibility of flying directly to their objectives, but are obliged to drag themselves along the ground, steering their way around the many obstacles by going over or around them. Ants have similar orientation problems finding their way in the frequently confusing system of passages and chambers in the nest. On the other hand, the ant workers' lack of wings is an important precondition for their continuing dominance on the ground, which is noted, above all, in the tropics.

The long life expectancy which has been demonstrated to exist among some ant species is a still less-appreciated feature of many ants. Whereas the lifespan of honey bee and also some ant species' workers such as *Monomorium pharaonis* is perhaps in the region of a few weeks or months, that of *Formica sanguinea* workers has been stated as being a maximum of five years, and that of *Formica fusca* as being a maximum of eight years (Stitz, 1939). Queens reach substantially greater ages. Queen bees have a life expectancy of between three and four years, whereas ant queens can greatly exceed 20 years-of-age. Appel has assured us (from Kutter, 1969) that a queen of the ant species *(Lasius niger)* reached what for insects is the biblical age of over 28 years. He captured the fertilized queen after her nuptial flight in August 1921, and 'cared' for her until she died in April 1950. The queen was able to lay fertilized eggs until the last years of her life. By contrast, ant males die a great deal sooner. They take part in the nuptial flight in the same year in which they hatch or the following year and die after mating.

1.2   Body structure and habits of ants

The typical ant colony consists of a queen and her female offspring, the majority sterile workers (*see* Fig. 1.1). While the workers procure food, feed the larvae, see to the building of the nest and defend the colony when under attack by taking the queen and the young to safety, the essential task of the queen is the laying of the eggs. She lays both fertilized and unfertilized eggs, from which hatch, as has already been stated, new queens or workers from fertilized eggs, and males from the unfertilized eggs. The majority of eggs produced by the queen develop as a rule into worker ants; male offspring and reproductive females result only at certain times of the year. The reproductive ants can usually be distinguished from the workers by their wings, whereas the workers possess none. Males and newly hatched full females do not

Fig. 1.1   Leptothorax nigriceps – queen (Q) with workers and brood (W).
(Photo – A. Buschinger)

usually remain long in the maternal nest. They often leave the nest
shortly after hatching to start out on the nuptial flight, along with large
numbers of reproductive ants from other colonies. The females are
fertilized during this nuptial flight, or directly afterwards. The males
have fulfilled their function with fertilization and die shortly afterwards.
The females, however, rid themselves of their wings and seek to found
a new colony, which only very few succeed in doing. The majority fall
to predators, or perish in other ways. Among some species, the fertilized
females are received by their mother colonies, or by foreign colonies of
the same species. Such colonies have several queens at their disposal,
which is described as polygyny (Reuter, 1913). In other colonies, where
only a single queen is tolerated, one refers to this as monogyny.
Pleometrosis and haplometrosis denote the occurrence of several queens
and a single queen respectively. By pleometrosis, one understands the
phenomenon whereby several fertilized females ally themselves for the
joint founding of a colony; in the case of haplometrosis the female
founding the colony is alone.

During the fertilization, a quantity of sperm is transferred from the
male to the female, and is stored in a 'sack', the 'receptaculum seminis'.
This quantity of sperm, which can be the result of one or several
fertilizations, generally lasts the female for her whole life. Fertilized
females of the monogynous species usually retire to a 'brood chamber'
and begin laying their eggs. The queen is obliged to nurture the larvae

of her first brood herself; subsequent broods are cared for by workers which have hatched in the meantime. There are a larger number of deviations from this 'basic type' of the life cycle and structure of an ant colony, which will be presented in more detail during the course of the book.

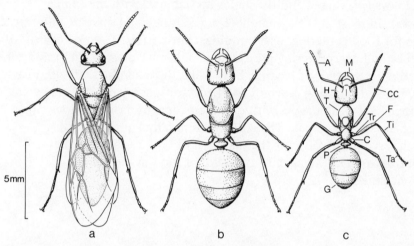

Fig. 1.2   (a) male; (b) female after shedding wings; (c) worker of the wood ant *Formica polyctena*. A. antennae; C. coxa; CC. cleaning comb; F. femur; G. gaster; H. head; M. mandible; P. petiolus; T. thorax; Ta. tarsal, Ti. tibia; Tr. trochanter (after von Frisch, 1974, adapted).

The structures of a female, a male and a worker are shown in Fig. 1.2, using the wood ant, *Formica polyctena,* as example of the subfamily Formicinae. As with all insects, one distinguishes in ants between the head and midsection, or thorax, consisting of the pro-, meso- and metathorax, and the posterior section or abdomen, which consists, in wood ants and many other species, of a joining section, or petiolus, and the remainder of the posterior section, the gaster. In ants of another subfamily, Myrmicinae, the joining section between the midsection and the rest of the abdomen consists of two parts, the front being the petiolus, the rear the postpetiolus. The legs consists essentially of five parts, the haunch (coxa), the thigh (trochanter), upper thigh (femur), lower thigh (tibia), and the foot sections (tarsals). The forelegs carry a cleaning comb between the tibia and the first tarsal.

The head of the ant carries the antennae, which consist of a long

initial section, the scapus, and several smaller sections which are called
funiculi or flagellae. The funiculum carries a number of different
sensory organs, which will be described in more detail later. Ants'
feelers and mouth parts, and this applies to insects in general, have
always been appendages to the body. The mouthparts of ants (*see* Fig.
1.3) consist of an upper lip (labrum), the outer jaws (mandibles), the
inner jaws (maxillae), and the under lip (labium) with the tounge
(glossa). In most ants the mandibles are serrated and serve as all-purpose
tools for food gathering, nest building, transportation, various forms of
defence and overpowering prey. The maxillae have an equally important
function in the task of feeding, but unlike the mandibles, are used for
grasping and breaking down of food and for sensory control. They are
equipped for this purpose with numerous, mainly taste-sensitive organs,
which are on the palpae. The labium is similarly equipped with feelers,
which serve principally for the taking-in of liquid food, and play an
important role in the passing-on of liquid food in the colony
(trophallaxis).

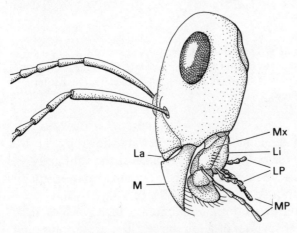

Fig. 1.3   Head of a *Formica lugubris* worker. La. labrum; Li. labium; LP. labial
palpae; M. mandible; MP. maxillary palpae; Mx. maxillae (after Sudd, 1967b,
adapted).

The function of the glands will be explained in more detail later. At
this point we only need to point out the functions and positions of the
most important glands (*see* Fig. 1.4). In the head region are the
mandible, propharynx (maxillary) and the postpharynx glands; the

labial glands also open into the head region, but are actually situated in the thorax. Also in the thorax are the metathoracal or metapleural glands, which similarly open into the thorax. The abdominal glands consist mainly of the poison and accessory or Dufour's glands. In the species of one subfamily, the Dolichoderinae, these abdominal glands have reformed considerably and developed into new glands, the anal and Pavan's glands.

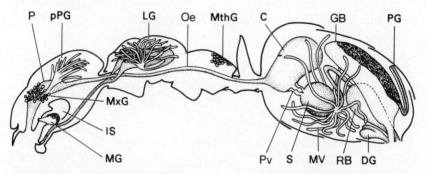

Fig. 1.4   Longitudinal section through a wood ant worker and representation of the intestinal tract and larger glands. C. crop; DG. Dufour's gland; IS. infrabuccal sack; LG. labial gland; MG. mandible gland; MthG. metathoracal gland; MxG. maxillary gland; MV. malpighi vessels (renal tubes); Oe. oesophagus; P. pharynx; PG. poison gland; pPG. postpharynx gland; Pv. proventricular valve; RB. rectal bladder; S. stomach (after Otto, 1962, adapted).

2 Classification and phylogeny of ants

Until 1968 (according to Bernard, 1968) 7600 recent (contemporary) species of ants forming the Formicidae family had been recorded. How large this family actually is we do not know, for even today new species can be recorded. Consequently, we can only estimate the number of all existing ant species which, according to Brown's calculations, lies somewhere between 12 000 and 14 000 species (from Wilson, 1971).

Whilst the ant fauna of Europe and North America has already been relatively well researched, we have reason to assume that there exists, especially in tropical areas, a large number of undiscovered species. This emerges from comparative figures collated by Kutter (1969). According to these, around 30% of the 110 ant species discovered up to the present time in Switzerland are social parasites (species which cohabit with ants of other species and live off them). The number, however, of all generally known socially parasitic ants discovered up to now is not larger than 200, which corresponds to roughly 2% of all the ant species known to exist. Of these, 56% come from the palearctic region, 27% from the nearctic region, 11% from the neotropics and 6% from all other regions, in other words Africa, India, China, Indonesia, Australia etc. Since few reasons can be discovered as to why the proportion of social parasites among the central European ants should be so high, except perhaps that there exists here a far more dense and homogeneous host population than in the tropics, these figures definitely and markedly point out the differences among the states of knowledge about ant fauna in different regions.

The ordering of large numbers of ant species into various subfamilies has been attempted many times, for example by Emery (1910-1925), Wheeler (1923), Clark (1951) and Brown (1954). If it were merely a question of dividing these ant species into groups so that they were in an order, then one classification would be as suitable as another. Since, however, in the modern system the classification of a group of organisms has to be a reflection of its phylogenic development, even the most widely known classification of Brown is at best tentative and as such only as reliable as the state of our knowledge about the phylogenic development of ants. Brown divides the ant family into nine subfamilies, the Myrmeciinae, Pseudomyrmecinae, Dolichoderinae, Formicinae, Ponerinae, Cerapachyinae, Dorylinae, Leptanillinae and the Myrmicinae.

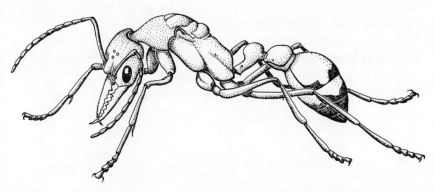

Fig. 2.1   Queen of *Myrmecia gulosa,* a primitive ant found throughout Australia (after Wilson, 1971, adapted).

The Myrmeciinae (Fig. 2.1) are found almost exclusively in Australia. As far as their anatomical structure and their behaviour are concerned, they are very primitive ants. By way of primitive features, they show a strong, well-developed sting, a full development of the maximum number of abdominal segments, labial and maxillary palpae and feelers. Other primitive features concern the anatomical structure of the legs and wings (among the sexual animals). Primitive characteristics of the males concern the structure of the genital apparatus. The various castes of the queens and workers are frequently linked by a series of intermediary forms. Among the habits of the Myrmeciines considered primitive is that queens founding a new colony go out hunting food at this time. In more developed groups of ants various alternatives have come into being in order to avoid such expeditions as would put the growing colony at risk. Adult Myrmeciines feed mainly on nectar, whilst the larvae are fed dead insects.

The Pseudomyrmecines live in the hollow branches of trees in the tropical regions of Asia, Africa and America and are compulsorily tree inhabitants. They are slender-looking animals and bear a definite resemblance, as do the Myrmeciines, to wasps, and can sting very uncomfortably. Their joining section between the thorax and abdomen is divided into two parts, the petiolus and postpetiolus. Their behaviour shows the ability to pass on liquid food from their crop directly to nest mates (trophallaxis). Another striking feature of this group is their development of a 'stridulation organ' (*see below*).

The Ponerinae, who also possess a well-developed sting, are an extremely heterogenous group, divided by Clark (after Brown, 1954) into a total of four distinct subfamilies, the Amblyoponinae, Discothyrinae, Odontomachinae and the Ponerinae proper. Brown (1954) regards these different groups (apart from the Ponerinae) as having the status only of species groups (Tribus). The group Amblyoponinae in particular includes species with very primitive

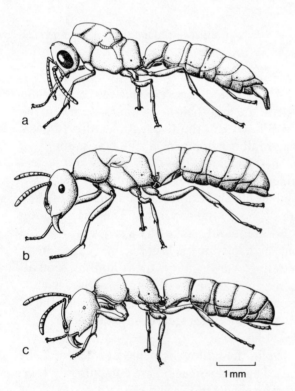

1 mm

Fig. 2.2   Amblyopone pluto, a primitive Ponerine; (a) male; (b) female; (c) worker (after Gotwald and Levieux, 1972, adapted).

characteristics. The structure of the wing veins and the joining section
between the abdomen and the thorax is very reminiscent of primitive
wasps of the group Tiphiidae, particularly *Anthobosca* and *Diamma*
(Brown, 1954). Figure 2.2 shows the various castes of *Amblyopone
pluto,* described only briefly by Gotwald and Levieux (1972). Here one
can clearly see that the petiolus is not yet separate from the abdomen.
The noticeable fact about the habits of the Ponerines is that they live
mostly above the ground and predatorily. Some species, particularly of
Onychomyrmex, recall the habits of army ants. Ponerines are
distributed for the most part in the tropics, whilst only very few species,
such as *Ponera coarctacta* and *Hypoponera punctatissima*, are present in
central Europe.

The Myrmicines take their name from the Myrmidones, those fearless
warriors whom Achilles mobilized against Troy and about whom the
saying is quoted that Zeus created them from an ant army. These
Myrmicinae, which are also widespread in Europe, are recognized by the
fact that a petiolus as well as a postpetiolus has formed (*see* Fig. 1.1).
This does not operate as a definitive characteristic among tropical ants,
since the Myrmicines of this region share the feature in common with
the Pseudomyrmecines and a few Dorylines. The striking characteristic
of this highly developed group is the tendency to feed on vegetable
food, the distinctive feature of leaf-cutter and harvester ants, which
belong to this group. Another tendency worthy of note, as well as being
indicative of the level of their development, is the regression of the sting.
We shall be investigating this interesting phenomenon in more detail
later. Among the Myrmicines, nearly half of the almost 500 species have
the use of an efficiently functioning sting.

The Dorylinae (Fig. 2.3) are army ants whose characteristic cycle of
settled and travelling phases is determined by the development of the
brood. All Dorylinae species have only one queen per colony, which is
always without wings and is, therefore, unable to undertake a nuptial
flight. The winged males differ very markedly in their external
appearance from the females and workers. Brown (1954) deduces that
the Dorylines are descended from the Ponerines, but does not rule out
the possibility that they, or a proportion of them, developed from the
Myrmicines. Some authors divide the Dorylines into two separate
subfamilies, the Dorylinae which are distributed throughout the Old
World, and the South American Ecitoninae (e.g. Markl, 1973a).

The Leptanillinae are minute ants — the average size of all *Leptanilla*
workers is under 1.5 mm — which live in small colonies and undertake

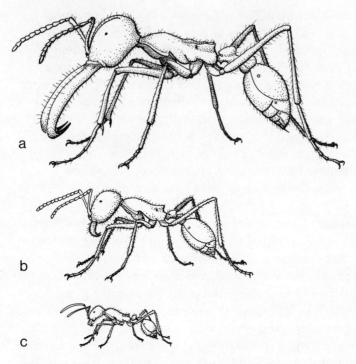

Fig. 2.3   Workers of the army ant species *Eciton burchelli.* The largest workers
(a) serve the colony as soldiers while the smallest workers (c) exclusively care for
the brood (after Topoff, 1972, adapted).

migrations when possible. The 'tiny ones' are usually not discoverable
by the usual methods of ant collecting, as a result of which we still
know only very little about them. Thus far, three genera have been
distinguished: *Leptanilla, Phaulomyrmex* and *Leptomisetes.* The
common characteristics of these three are that the workers have no eyes
and the maxillary and labial palps are reduced. The females have no
wings and, like the workers, have no eyes, or only very small ones. The
males have short, unserrated mandibles and have wings. The second pair
of wings has no veins, and their first pair only a few (Fig. 2.4). The
external sexual apparatus of the male is broad and cannot be drawn in.
The larvae of the Leptanillines are particularly characteristic: like the
Leptomisetes, all known Leptanilline larvae have a spoon-like appendage
fixed to the ventral side of the front portion of the body, upon which
food is placed and conveyed into the larva's mouth by their
continuously grating mandibles. (Kutter, 1948).

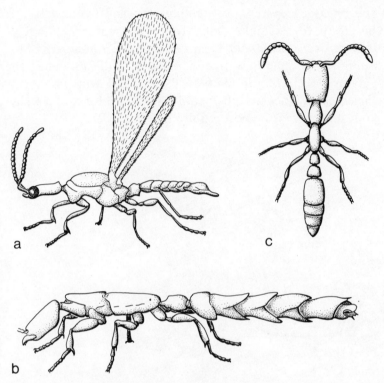

Fig. 2.4   (a) male of *Leptanilla miniscula;* (b) female; (c) worker from
Leptanilla revelieri (after Wheeler, 1910, adapted).

The status of the Cerapachyinae as a distinct subfamily is still not
secure. It is quite possible that they will be re-integrated into the
Ponerinae as a species group (Tribus), as they already have been by
Brown and Nutting (1950). Brown treats them as a separate subfamily,
although he is still not completely convinced of his classification. As far
as we know of these rare groups of ants, all Cerapachyines feed on other
ants which they mostly steal as larvae or pupae. Some species, like
Cerapachys opaca (New Guinea) and *Phyracaces cohici* (New Caledonia)
also feed on fully grown animals (imagos); we still do not know how
selective they are in this, if they have specialized in particular types or
not (Wilson, 1958c). A few of the Cerapachyines go hunting in groups,
as will be described in more detail later.

The Dolichoderinae and the Formicinae count among the most
developed ants. Whilst the Formicinae are represented by a large
number of types in central Europe and are widely distributed – all

wood ants (*Formica*), carpenter ants (*Camponotus*) and *Lasius* ants belong to this group, for example — most Dolichoderines, except a few such as *Dolichoderus quadripunctatus* and *Tapinoma erraticum* in Europe, are only to be found in warmer regions. In both subfamilies the stinging apparatus is regressed. In the typical case, the Formicines are easily recognized by the fact that the petiolus takes the form of a scale or a single node and there is no postpetiolus (*see* Fig. 1.2).

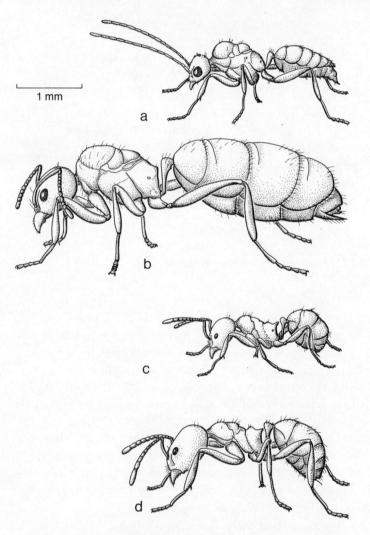

Fig. 2.5   Aneuretus simoni. (a) male; (b) female; (c) worker; (d) soldier (after Wilson *et al.,* 1956, adapted).

In addition to these new subfamilies, the Aneuretinae are also frequently mentioned nowadays as another subfamily. This group, with the *Aneuretus* and the extinct genera *Proaneuretus, Paraneuretus* and *Mianeuretus,* is of special interest, since, from the point of view of historical development, it stands between the primitive Myrmeciines and the more developed Dolichoderines. At all events until a few years ago we knew less about this ant group than almost any other. Until 1955 we only knew of a single type of Aneuretine found in three localities in Ceylon. The outcome of these discoveries consisted of not more than five or six specimens, which were preserved in various museums throughout the world. Only when Wilson in 1955 discovered a total of 20 nests in Ceylon and could, as a result of these, ascertain the various stages of development and the castes, did it become possible to investigate the internal anatomy of these animals, their ecology and habits (Wilson *et al.,* 1956). The interstitial position of the Aneuretines between the Myrmeciines and the Dolichoderines was ascertained through comparative examinations of the mouthparts (Gotwald, 1970) and the poison apparatus (Hermann, 1968). According to the field observations of Wilson *Aneuretus* is mostly predatory and nocturnally active; their larvae receive the captured animals whole, in the manner of many primitive types. In artificial nests the workers also consume honeydew and pass it on to their nest mates.

Finally an eleventh, now of course extinct, subfamily of ants, the Sphecomyrminae. Until the discovery of the first, and thus far only, representative of this subfamily, which was found embedded in amber, the fossil discoveries of ants extended in time from about 40-50 million years ago (Wheeler, 1914). Wilson and Taylor described in 1964 a fossilized colony of a species of *Oecophylla* from the early Miocene period in Kenya, which contained the various worker subcastes as they are found today in this genus. This find demonstrates not only the great age of this genus, but also shows that a specific social system has been maintained for at least 30 million years. Of course this find did not supply an answer to the question about the forbears of ants and they searched long and in vain for earlier specimens. Two rock collectors, Mr and Mrs Frey, found a piece of amber in the friable surface chalk of a cliff at Cliffwood in New Jersey. Two fragments of this piece contained insects, beneath which were two well-preserved ant workers (Fig. 2.6), which were named *Sphecomyrma freyi* (Wilson *et al.,* 1967). The age of this find was specified as 100 million years. This meant that a well-preserved fossil had been discovered which was at least twice as old as the oldest previous discoveries. With its age of 100 million years, *Sphecomyrma freyi* was not only the first known ant from the Mesozoic age, but was also the first mesozoic insect of any description to be preserved.

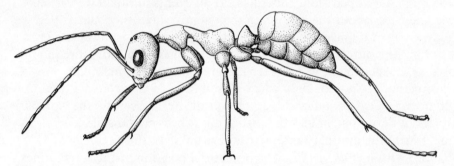

Fig. 2.6   *Sphecomyrma freyi,* the first known mesozoic ant (after Wilson, Carpenter and Brown, 1967, adapted).

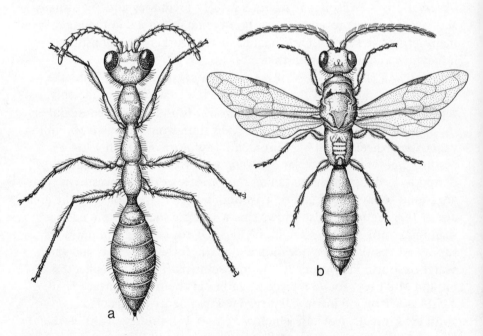

Fig. 2.7   (a) female; (b) male of the wasp *Methocha fimbricornis,* belonging to the Tiphiides and widespread in the Philippines (after Wilson, 1971; from F.X. Williams, 1919).

The discovery of *Sphecomyrma* is a very important indication that ants may well be descended from wasps of the Tiphiidae group, particularly from those which most closely resemble species of *Methocha* existing today. This refutes the hypothesis of Morley (1938), which refers to Emery (1896) and Forel (1921-1923), namely that the Mutillidae were the ancestors of ants. Wheeler (1928) and Brown and Nutting (1950) on the other hand, already regarded the Tiphiides as the most probable ancestors of ants. Wheeler explicitly excluded *Methocha* of course with the Mutillidae as ancestors of ants, pointing out the absence of wings in the females.

As typically ant-like features, *Sphecomyrma* exhibits above all the metathoracic gland, a characteristic common to nearly all ants, and the ant-like 'waist' with completely separate petiolus. On the other hand, the antennae of *Sphecomyrma* are wasp-like, as are their very short and poorly serrated mandibles. From the relatively small differences between *Sphecomyrma* and the Tiphiides Wilson *et al.,* (1967) presume that the social evolution of ants took place in a relatively short time.

Sphecomyrma has closer affinities with the Myrmeciines than the Amblyopones or other Ponerine species, from which both were previously believed to have descended.

There are two different representations of phylogeny within the ant family. According to Wheeler (1928) the Ponerines are the earliest group, from which developed the rest of the subfamilies. According to Brown (1954), however, ants split off into two separate groups early on in their development, the two groups being the 'poneroid' and the 'myrmecioid' complexes of subfamilies. In the first group, according to Brown, the Leptanillines, Dorylines and Myrmicines developed, in the second the Myrmeciines, the Sphecomyrmines, Pseudomyrmecines, Aneuretines, Dolichoderines and Formicines respectively. Figure 2.8 represents a hypothetical ant family tree, based on Brown's ideas. According to the research of Markl (1973a) however, the Pseudomyrmecines fall outside the myrmecioid complex, since they were the only group in this group to form a stridulation organ, which is otherwise present only in representatives of the poneroid complex.

Wheeler's hypothesis is based on that of Robertson (1968), who compared the poison apparatus among some Hymenoptera groups. The hypothesis of Brown on the other hand was based on the investigations of Eisner (1957; *compare also* Eisner and Wilson, 1952 and Eisner and Brown, 1958), who compared the proventriculus of different ant groups. The proventriculus links the two central chambers of the gut, the crop and the intestine among Hymenoptera and many other insects (*compare* Fig. 1.4). Whilst digestion takes place in the intestine, among social insects the crop has developed into a 'social stomach', in which food is stored and then passed on to nest mates. The proventriculus consists of a muscular portion, round in shape, the bulbus, which is joined to the crop by means of an opening in the form of a cross, and to the intestine by a slender, extended valve. This has the function of pumping essentially liquid food from the crop to the intestine.

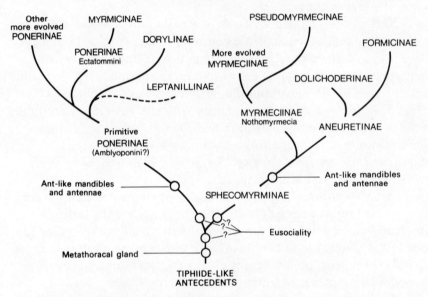

Fig. 2.8   Hypothetical family tree of ants, showing the descent connections of the various subfamilies (after Wilson, Carpenter and Brown, 1967).

Gotwald (1969) has investigated the phylogenic relationship among ant subfamilies by comparison of the mouthparts. Amongst other things, this revealed the close relationship between the Cerapachyines and the Ponerines, whose mouthparts differ in only one characteristic. Gotwald, however, hesitated to place the Cerapachyines with the Ponerines. The work of Gotwald brought no clarification as to a decision between the hypotheses of Wheeler (1928) and Brown (1954). His comparison of the mouthparts reflected neither the great chasm lying between the Myrmeciines and the Ponerines, which Brown had challenged, nor did it adequately make a clear connection between the

Ponerines and the Formicines to support Wheeler's hypothesis. We can see, therefore, that further comparative research is required before we are able to reach a definitive decision about the phylogeny of ants. How far comparison of the number of chromosomes (Crozier, 1970; Hung *et al.*, 1972), comparison of larval forms (Wheeler and Wheeler, 1970) or comparative behavioural research (e.g. Brown and Wilson, 1959) will lead us remains to be seen. In any case, clarification of the phylogeny of ants rests on the hope of future fossil finds. Further discoveries are expected, for example, in Canadian amber to be found in the river basin of the Saskatchewan river (Mac Alpine and Martin, 1966). In the amber from the surface chalk interesting fossils of other insects have already been discovered, and it is possible that here also, further specimens of mesozoic ants will be found. A piece of amber was likewise discovered in the chalk of the arctic coastal plain of Alaska (Hurd *et al.*, 1956), the age of which is not known exactly. Finally, in Lebanon a fossilized pine was discovered in the under chalk, in which an ant and other fossils were found, but which has not yet been described (Hennig, 1969). This highly interesting ant discovery in under chalk shows that ants existed before the Cretacean period, i.e. more than 135 million years ago.

3 The sensory organs of ants

Differentiated and numerous sensory organs are important for the social life of ants, since the signals of the social partner can also consist of a variety of optical, chemical and mechanical signals (*see* Chapter 5). Sensory organs are, moreover, essential for the orientation of ants, the subject of Chapter 4.

3.1 Sensory organs of the feelers

The feelers of ants consist of an extended initial section (scapus) and up to 12 shorter sections, which together form the funiculus. The first section of the funiculus, which is called the pedicellus, contains a mechanical receptor organ called the 'Johnston Organ'. Other sensory organs lie in large numbers on the surface of the funiculus. The immediately noticeable manoeuvreability of ants' feelers is essentially achieved by two joints, a ball and socket joint between the scapus and the head (*see* Fig. 3.4) and a hinge joint between the scapus and the pedicellus. These jointed sections can be moved in opposite directions powered by muscles. The rest of the feeler segments contain no muscles and the limited manoeuvreability of these segments is a result of changes of pressure in the blood plasma (haemolymph).

The sensory organs of the feelers of ants had already been researched over 100 years ago by Forel (1874); he distinguished between five different types: 1. Leydig's cones or olfactory bulbs; 2. fine, pale and pointed sensory hairs; 3. long fine bristles which are bent round close to the joint and run parallel to the antenna surface as a result; 4. organs shaped like champagne corks; and 5. bottle-shaped organs.

With his designations 'olfactory bulbs' and 'sensory hairs', Forel seems to have anticipated events, since until that time there had been no exact research into the functions of these sensory organs. We have become rather more cautious since then, describing the 'olfactory bulbs' as sensilla basiconica and the 'sensory hairs' as sensilla trichodea

(Schenk, 1903). 'Sensillae trichodeae', which are joined to the surface of the feelers by an articulated membrane, have been given another name by Schneider and Kaissling (1957) — 'sensilla chaetica'. The bent hairs (Forel) are nowadays usually described as 'sensilla trichodea curvata' (Krausse, 1907), the champagne cork-shaped organs as 'sensilla coeloconica' and the bottle-shaped organs as 'sensilla ampullacea'.

The various types of sensory organs on the antennae of ants and their distribution on the feelers were investigated in more depth later by Dumpert (1972b). On the antennae of *Lasius fuliginosus* apart from the sensilla trichodea curvata, sensilla basiconica, sensilla coelonica and sensilla ampullacea described by Forel, there are sensilla chaetica and sensilla campaniformia, which are grouped together in Fig. 3.1. These various types of sensillae are all found on the antenna, particularly on the pointed segments. The rear segments, especially the scapus, carry sensillae chaeticae almost exclusively. On the antennae of other types of ants and also various subfamilies, as with *Myrmica laevinodis* (Myrmicine) and *Dolichoderus quadripunctatus* (Dolichoderine), the same sensillary types are found in the same distribution as with *Lasius fuliginosis*. A sensilla type described by Kürschner (1969) as having an 'embryonic incision' had, however, little corroboration (Dumpert, 1972b), nor did a 'new type of sensilla' described by Masson and his colleagues (1972) regarding *Campanotus vagus* (Walter, reported personally).

On the approximately 2 mm long funiculus of a *Lasius fuliginosus* worker there are altogether about 3000 sensillae, of which the majority are sensilla chaetica. After a large gap, the next most numerous are the sensilla trichodea curvata, then the sensilla trichodea, the sensilla basiconica and finally the rest of the sensillary types, which occur only sporadically. The males of *Lasius fuliginosus* have fewer sensillae, corresponding to the smaller surface area of the funiculus; there occur, however, the same sensillary types as with the workers.

Ant species possessing only poorly developed eyes, or none at all, such as the majority of Dorylines, have more numerous and more dense sensillae on their feelers than other ants with well developed eyes (Weiler, 1936; Stumper, 1955a).

If one compares the various sensillae to be found on ant antennae, the sensilla coeloconica, and more especially the sensilla ampullacea must be excluded, since they are positioned under the surface of the antennae. In the case of the sensilla coeloconica, a conical shaped body is situated relatively close to the underside of the surface inside a cavity;

with the sensilla ampullacea, a somewhat long sensory hair is usually situated rather further beneath the surface of the antenna and is generally connected with it by a canal (Fig. 3.2). The opening of the cavity of the sensilla coeloconica and of the canal leading to the sensilla ampullacea has a diameter of only 0.5 μm. These structures, originally discovered by Forel, have aroused the particular interest of many authors (Prelinger, 1940; Jaisson, 1969, 1970; Masson and Friggi, 1971a,b), and there has been much puzzling and speculation about their function.

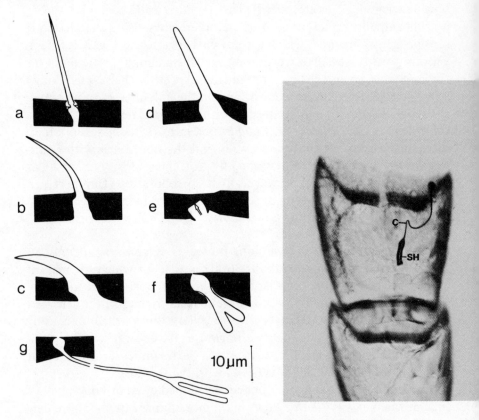

Fig. 3.1   Diagram of the cuticular structure of antenna sensillae of *Lasius fuliginosus:* (a) sensilla chaetica; (b) s. trichodea; (c) s. trichodea curvata; (d) s. basiconica; (e) s. campaniformia; (f) s. coeloconica; (g) s. ampullacea (from Dumpert, 1972b).

Fig. 3.2   Second and part of third antenna segment of a *Lasius fuliginosus* worker with sensilla ampullacea. C. canal, showing the connection to antenna surface; SH. sensory hair (from Dumpert, 1972b).

Demonstrating the function of the sensory organs on the feelers of ants is impeded, as with most other insects, by the fact that the various types of sensilla are mixed together on the feelers. It is, therefore, not possible to remove individual specimens of sensilla by amputation of sections of the feeler or by other means, and prove by its retention which particular sensory function emerges. Electrophysiology provides the only certain method: in this way the action potential of the sensory cells of an individual sensilla can be diverted by means of extremely fine electrodes and amplified sufficiently to be perceptible on an oscilloscope. If one applies various stimuli to the sensillae, such as, for example, odours, carbon dioxide, moisture, light, heat, cold and so forth, then one is able to ascertain from changes in frequency of the spike potentials, what the sensillae are sensitive to, and thereby determine their function.

With this method, which has also proved fruitful in determining the function of other insect sensillae, Dumpert was able to demonstrate with *Lasius fuliginosus* that the sensilla trichodea curvata, the sensilla trichodea and the sensilla basiconica are odour receptors since at least 30 sensory cells of the sensilla trichodea curvata respond in a particular manner to specific groups of chemical substances (ketones, fatty acids, alcohol etc.). One of the sensory cells reacts specifically to the alarm substance of this species (Undecane). The sensillae campaniformiae react to changes in tension of the cuticula and the sensillae chaeticae to the mechanical inflection of the bristles and serve very probably as touch organs. Morphological research (Dumpert, 1972b) showed that in addition to the mechano-receptive sensory cells ending at the base of the sensilla, there are among certain sensillae chaeticae further sensory cells which extend to the tip of the hair. Apart from this it was shown that among at least some sensillae chaeticae stain solution infiltrates from the tip into the sensory bristle, which leads us to conclude that they have an opening here, which is typical for taste bristles.

The functions of the sensilla coeloconica and sensilla ampullacea, deviating in their structure from the other sensillae, is more difficult to determine, as they are only visible by other means from the outside which prevents electrophysiological examination. On the basis of less successful deductions, however, one can say that the sensillae ampullaceae are very probably purely carbon dioxide receptors, whilst the sensillae coeloconicae among others respond to rises in temperature. Whether or not there are other moisture- and cold-sensitive receptors under the three sensory cells of this sensilla coeloconica or under the sensory cells of other sensilla types, as has already been demonstrated on the feelers of other insects, remains to be seen.

It has already been proved in behavioural research that ants are able to distinguish different temperatures, which is important for them. They are thereby in a position to maintain themselves in places which best correspond to their optimum temperature. The development of the larvae is also affected by temperature. The best develop between 23 and 29°C (Steiner, 1925), and are transported by the workers to those places in the nest which most closely correspond to this temperature range. Herter (1924) has been able to show that wood ant workers are able to distinguish temperature differences of between 1 and 4°C.

It was similarly established through behavioural research that taste sensitive organs are to be found on ant feelers. In order to prove this, Schmidt (1938) fixed workers of *Myrmica ruginodis* to a base with their backs downwards and held various solids and liquids to the feelers (water and sugar solution). This showed that the ants stretched forward their underlip and tongue on each contact of their feelers with a liquid. For finer distinctions, these spontaneous reactions of the animals were insufficient: one could not ascertain by this method whether they were able to distinguish between various taste stimuli. In order to determine this, and indeed it depended on this connection, Schmidt either rewarded or punished the reaction to various stimuli. If the ants stretched out their tongues at the sugar solution, they were given some of it to drink. If, however, they stretched out their tongues on contact of the feelers with water, they were given a quinine solution, which produced a strong defensive reaction from the animals. After these experiments had been repeated many times, the ants learned to react differently to water and sugar solution, thus demonstrating that they were able to distinguish the two liquids with their feelers. By means of further experiment, Schmidt was, in addition, able to exclude the possibility of the animals distinguishing the different liquids with the aid of mechano-receptors distinguishing their viscosity. This showed that there are taste sensitive organs on the feelers of ants. In contrast to bees and flies, however, there are no taste sensitive organs on ants's legs.

The existence of light receptors on the feelers of carpenter ants (*Camponotus*) has been determined by the research of Hug (1960) and Martinsen and Kimeldorf (1972). Hug treated the ants with X-rays, and observed that the animals reacted with movement of the feelers at 0.05 rad/s, and with increased mobility 0.5 rad/s. Martinsen and Kimeldorf were able to deduce the total potential of the antennae, after they had stimulated them with β rays of 62 mrad/s. This stimulus intensity is so great, however – the tolerance level of the population, by way of comparison, lies at 0.1 mrad/h for β and γ rays – that what we see here is very probably a non-physiological effect. Research by Cadwell (1973) and Brower (1966) reveals the extraordinary resistance of ants to X-rays: young colonies of American harvester ants, *Pogonomyrmex badius,* were subjected to continual radiation of 18 rad/s without the colonies being dispersed or noticeably disturbed in their development.

A further sensory organ, situated in the feelers of ants and many other insects, is Johnston's organ. It consists of numerous sensory cells which lie within the pedicellus together with auxiliary structures. They are so arranged as to register the flexing of the funiculi over the pedicellus. This organ was more closely examined by Masson and Gabouriaut (1973) with an electron microscope, using *Camponotus vagus.* There are over 500 sensory cells contained here.

According to Vowles (1954a), Johnston's organ serves ants as a gravity sensor, the funiculi being bent by gravity, and this deflection registered by the organ. Vowles determined this by experiments with *Myrmica rubra* and *Myrmica ruginodis,* in which he stuck iron filings to the abdomen, thorax, head and feelers, making them move in such a way as to orient themselves to gravity. By using a magnet, he was able to bend the jointed segments, and ascertained that the animals only took up a new direction when the flagellum was bent opposite the scapus.

Vowles's interpretation, however, could not be supported. In the first place, one should say that it would be exceedingly unsatisfactory to place gravity sensors exclusively in the feelers, where a mere gust of wind could mislead the ants in their gravity orientation (Bückmann, 1962). Apart from this, it has been demonstrated with *Myrmica rubra* that they are also able to manoeuvre by gravity orientation without feelers (Markl, 1962). It is thus unequivocally clear that Johnston's organ is not necessary for the gravity orientation of ants. The research of Vowles, however, leads us to agree, as also emerges from the research of Markl (1962), that Johnston's organ does serve to register air currents.

3.2   Sense organs of equilibrium

Although ants, like all insects, do not have statocysts, they are nevertheless well able to orient themselves by gravity. Such orientation undoubtedly plays an important role in the nest in the dark, particularly in large nests, with their abundance of passages and chambers. The structure of the nests themselves shows obvious orientation of the passages by gravity in many respects.

The precision with which ants are able to orient themselves by gravity emerged when gravity oriented *Formica polyctena* workers covered a distance 100 mm long with a path of 104 mm. Ants thus move almost in a straight line; in isolated cases completely straight tracks were found. Wood ant workers can recognize gravity orientation even by a sloping angle of 3.5° (Markl, 1962).

Nothing precise was known for a long time about the sense organs which made such orientation possible. In 1959 Lindauer and Nedel were able to demonstrate definitively in the case of honey bees that bristle areas on the throat and midsection function as sensory organs for equilibrium. In view of the close relationship between bees and ants, it seemed desirable to look for bristle areas in ants also and to determine in a given case whether they are responsible for gravity reception.

In fact Markl found in the subfamilies he examined, the Formicines, Myrmicines, Ponerines, Dolichoderines and Dorylines, between 24 and 26 bristle areas, which were partly paired off. All these bristle areas lie on joints, as one can see from the specimen of the wood ant *Formica polyctena* in Fig. 3.3. It is a common feature of all the bristle areas that they consist of dense sensory bristles of approximately equal length. One can see under the microscope that the areas carrying bristles are lighter in colour than the surrounding cuticle. One can conclude from this, either that they are thinner, or that they contain less pigment. The transmitting fibres of all sensory cells lead to the central nervous system in a common nerve fibre.

The bristle areas are so arranged that they are stimulated by another jucture according to the joint on which they are found. On the bristle field of the neck joint the wall of the head capsule, which surrounds the occipital foramen, serves as counter surface. As in this case, so in all others the bristles are stimulated by being bent back with varying force according to the position of the joint. They thus serve as 'proprioreceptors', indicating the previous position of the joint in question.

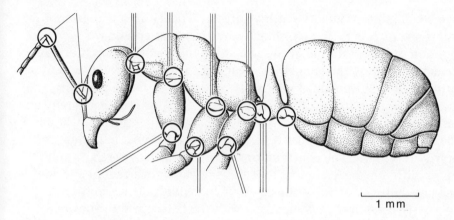

Fig. 3.3   The position of the various proprioreceptive bristle areas of a wood ant *Formica polyctena* worker (after Markl, 1962, adapted).

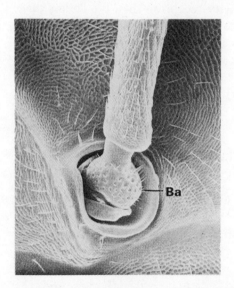

Fig. 3.4   Scanning electron micrograph of the bristle area (ba) on the ball and socket joint between the feeler and the head capsule of *Lasius niger* (Photo – W. Gnatzy).

Markl (1962) bypassed the use of the bristle areas by fixing the joints or singeing off the bristles. It was shown in a special test that in *Formica polyctena,* the bristle areas serve the feeler, neck, petiolus, abdominal and coxal joints as gravity sensitive organs. Gravity affects the various joints and directs them according to the gravity orientation. The position of the joints and at the same time the gravity orientation can be determined by the animals by means of the proprioceptive joint bristles. Of the five named joint systems, each is self-sufficient as a gravity receptor. If one of the bristle areas fails, however, it has only an insignificant effect upon gravity reception. The paired bristle field on the neck joint does the most work, and is thus the most important for gravity reception. The contributions of other bristle areas show only minor differences. According to Markl's research, one cannot exclude the possibility of other bristle areas, or Johnston's organ, contributing to gravity reception; in comparison with the five named bristle areas, however, these contributions must certainly be very small.

The proportion of the various bristle areas allotted to gravity reception is in any case not the same with all ant species. The few comparative researches of Markl show that the feelers of the harvester ant *Messor barbarus* effect more gravity reception than the feelers of *Formica polyctena*. It would seem to be the case, however, that the responses of the various organs are centrally calculated. In this manner one avoids the chance of something other than a mechanical effect dependent on gravity, upon a joint concerned with gravity reception, leading to a misapprehension. False information from only a single joint does not lead to a miscalculation of the animal's gravity course.

3.3   Sensory organs for hearing and vibration

Both hearing and vibration sensitive organs react to mechanical oscillations; the hearing organs to sound waves, mostly atmospheric noise (rhythmic changes in pressure), the vibration receptors to aperiodical oscillation, mostly of solid or liquid bodies. In exceptional cases the vibrations of solid objects may also be perceived as sound. An example of this is the sound one hears if one places a tuning fork which has been struck on to the teeth. In this case, no atmospheric sound results, but one nevertheless hears the vibrations of the tuning fork as a tone. This example clearly shows that one cannot make any statements about the subjective perceptual quality mediated by specific sensory organs in the case of organisms other than human beings. Whether, therefore, ants in a position to perceive sound experience a sensation corresponding to our hearing, is something we can comment as little about as the subjective sensation of ants perceiving vibrations. Neither sensory organ – and this applies equally for the other sense organs – is characterized by its sensations, so much as by its adequate stimulus. By way of the delineation of the hearing and vibration receptors, Markl puts forward the suggestion that all corresponding sense organs operating in the near distance should be described as vibration organs, and all sense organs reacting to sound waves in the far distance as hearing organs.

The search for the respective hearing and vibration organs was desirable, because some species produced sound with a 'stridulation organ', while others made a tapping signal with their posterior segment. A stridulation organ is to be found on all Pseudomyrmecines, 48% of all researched Ponerines and 83% of Myrmicines researched (Markl, 1973a). It consists of a ribbed area, which looks very similar to a washing board, and a stridulating edge, which can be moved over the ribbed area. In Myrmicines and Pseudomyrmecines the ribbed area lies on the outwardly curved anterior end of the abdomen, the sharp rear end of the postpetiolus serving as the stridulating edge (Figs 3.5; 3.6). During the stridulating movement, the posterior segment is turned on to its transverse axis and the stridulator is simultaneously pressed against the ribbed area. The result is a sound, whose frequency depends on the distance between the ribs and the speed with which the stridulator and the ribbed area are moved against each other. Its sound-energy, on the other hand, depends on the size of the sound-emitting surface, and thus on the size of the ant.

These stridulating organs of certain ant species have been known for a relatively long time. According to Autrum (1936), Swinton had already observed (1878/1879) that *Myrmica ruginodis* moved the rear portion of its body from side to side every so often, and as a result discovered the stridulation organ. Adlerz (1886) made similar observations with *Leptothorax tuberum* and *Harpagoxenus sublaevis*, but was able to perceive as little in the way of sound as Swinton. Although the noises produced by *Myrmica* are not audible to humans without instruments, the first success with the detection of stridulation noise was by Mc Cook (1880) with *Myrmecocystus melliger* from Texas, by Wheeler (1910) with the harvester *Pogonomyrmex barbatus molefaciens* and the females of the leaf-cutter ant *Atta fervens*, and by Krausse-Heldrungen (1910) with the harvester species *Messor barbarus* and *Messor structor*.

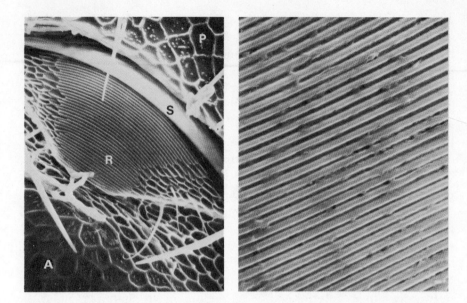

Fig. 3.5   Scanning electronmicrograph of the stridulation organ of a small (minima) worker of the leaf-cutting ant *Atta cephalotes* (enlarged 1470 x).
A. abdomen; P. postpetiolus; R. ribbed area; S. stridulator (Photo — W. Gnatzy).

Fig. 3.6   Section of the ribbed area of the stridulation organ of a major worker of the leaf-cutter ant *Atta cephalotes* (enlarged 3260 x) (Photo — W. Gnatzy).

Among many species of stridulating ants the noise produced can barely be heard, if at all; there are numerous reasons for this. The sounds either lie at a frequency level which is outside the range of human hearing, or the resonance is so little, that the sound is too quiet for our ears. Markl (1965, 1966, 1968) researched both the frequency range and the resonance of stridulation noise which was audible from a short distance away, and came to the following conclusion: the spectrum of sound by this method reaches from 5000 to 80 000 Hz, whilst the maximum resonance lies in the supersonic area between 20 000 and 60 000 Hz. The level of a large worker stridulating at a distance of 0.5 cm amounts to 74 dB, at a distance of 2 cm 66 dB and at a distance of 5 cm, 58 dB. The greater part of the ant's output thus lies in the inaudible supersonic range, whilst only 10% lies within hearing range between 5000 and 15 000 Hz. The human ear requires something like 30 dB greater output in this frequency range than in its optimum hearing frequency of 1000 Hz. Since the noise of an *Atta cephalotes* worker stridulating at close range is 40 dB over the hearing threshold of humans, it is thus audible. The smaller leaf-cutter ants and similar small ants are only audible with much greater effort, or not at all. Among other species, such as *Acromyrmex octospinosus*, which is also included among the leaf-cutters, a large proportion of the resonance output seems, by contrast, to be within the hearing range of humans (Kermarrec *et al.*, 1976).

It is not important for ants that we are able to hear their stridulation sound. Much more important for the function of the stridulation organ is whether the ants are in a position to hear their own sounds. Autrum (1936) investigated this question with *Myrmica ruginodis* and *Myrmica rubra*, and found that the workers stridulate when they are stopped, when they are heaped on top of one another, when exposed to alcohol or ether fumes, when they are injured by heat and occasionally whilst eating. The *Myrmica* workers stridulate only for a short time when eating, and then only when there are a number of ants crowded at the feeding area. The *Myrmica rubra* and *Myrmica ruginodis* workers in any case showed no reaction to their own stridulation noise, even when this was transmitted to the ants through the underlying surface. Behavioural research thus brought no indication that ants respond to their stridulation noise with a specific action, despite the fact that they are able to perceive atmospheric sound, as Autrum has demonstrated.

With *Megaponera foetens* belonging to the Ponerines, Collart (1925) found that groups of workers were led to attack a termite colony by stridulating nest mates. Levieux (1966), however, was unable to confirm this observation.

Markl discovered the first certain function of stridulation for the social life of ants, whilst observing South American leaf-cutter ants under natural conditions (Markl, 1967). As Autrum had done with the *Myrmica*, he found that the leaf-cutters usually stridulate when they are unable to move freely. This is the case when one takes hold of them with fingers or pincers, when they are buried underground, when they are surrounded in a fight with other ants and held fast and, finally, when they fall into water or are caught in a spider's web. Workers and reproductive animals are alike in this respect. In contrast with Autrum, however, Markl found a distinct behavioural response among the ants he researched to the stridulation signal. The workers of *Atta cephalotes* were drawn to the stridulation signal of the same species only when the signal was transmitted by underground vibration. One was able to observe no reaction on the part of the leaf-cutter ants to the atmospheric noise resulting from stridulation. If, however, a stridulating ant lay buried under a layer of earth, the attracted workers began to dig at the point where the stridulation intensity was greatest. Nevertheless, even a large worker could not lie deeper than 5 cm in order to continue attracting them, and not deeper than 3 cm to bring about a digging reaction. These observations show that stridulation acts as an emergency signal for buried nest mates among *Atta cephalotes.*

Under natural conditions, however, single workers are very seldom dug out, whilst groups of ten are dug out regularly. This is significant, inasmuch as in a colony of several million leaf-cutters, a single worker does not play a great part. The loss involved in digging them out and freeing them bears no economic relation to what the colony gains by their being saved. It is a different case, however, when larger groups of workers and above all the queen are buried. Here the rescue of the trapped ants is extraordinarily important for the continuance of the colony. When many ants are trapped, then not only is the intensity of stridulation greater than with individual trapped animals, but so also is the length of the alarm signal. Whilst individual ant workers seldom stridulate continually for more than 5 min., Markl was able to record the lively stridulation of one group of trapped ants, still continuing after 16 h.

The danger of becoming trapped is quite substantial for leaf-cutter ants. The nests of large colonies are buried up to 5 m deep in porous ground, and consist of several hundred metres of passages and numerous chambers. A heavy tropical rainfall can do much damage here. Similarly, however, the natural enemies of the leaf-cutter ant, such as anteaters, can devastate nests and bury portions of the colony.

With another stridulating ant species, the harvester ant *Pogonomyrmex occidentalis* found throughout America, Spangler (1967, 1974) was able to distinguish three different stridulation signals according to the alternation of different stridulation intensities and different levels of stridulation noise. It is possible that more information is contained in these different signals, and thus that stridulation does not exclusively and under all circumstances indicate an emergency alarm for trapped nest mates. This, however, needs to be researched in more depth.

Since the leaf-cutter ants only react to stridulation signals transmitted by underground vibration, there must obviously be vibration receptors in their legs, with which they perceive the signal. Among ants and other insects, such vibration sensory organs have been known for some considerable time as the 'subgenual organs', which are situated in the area of the tibia, immediately below the knee joint.

The structure of the subgenual organs has already been researched by Schön (1911). The organ contains between 10 and 20 sensory cells, which lie in the middle of the bone. On one side the organ grows attached to the wall of the tibia, on the other the transmitting fibres of the sensory cells form the subgenual nerve, to which the organ itself is attached. Autrum and Schneider (1948) have investigated the sensitivity of this organ, and discovered that among carpenter ants (*Camponotus*), it reacted most sensitively between 1.5 and 3 kHz and responded here to amplitudes of 1.8×10^{-6} cm. With the leaf-cutter ants (*Atta cephalotes*) Markl (1970) discovered that they were able to perceive underground sinary vibrations of between 0.05 and 4 kHz. The optimal functioning level of the receptors — if one regards them as amplitude receivers — is between 1 and 3 kHz, as was the case with *Camponotus*. If one regards them as accelerator receivers, however, then their most sensitive range lies between 0.1 and 2 kHz. There are thus variations of sensitivity in these receptors, as much between those of different large workers as there are between different legs of the same worker. The receptors of the front legs are four to five times more sensitive than those of the middle and hind legs; the receptors of small workers are more sensitive than those of larger workers. The lowest amplitude threshold lies, according to the calculations of Markl, at 1.3×10^{-7}, one tenth of a power below the threshold of the *Camponotus* subgenual organ as calculated by Autrum and Schneider.

It is interesting to note, however, that Markl was investigating a different sensory organ with *Atta cephalotes* from that investigated by Autrum and Schneider with *Camponotus*. Although leaf-cutters possess a subgenual organ, they receive ground vibrations chiefly through three groups of sensillae campaniformiae, which are situated between the femur and the trochanter (Figs 3.7; 3.8). This is surprising, inasmuch as until now we have not known of sensilla campaniformia functioning as vibration receptors with any other insect. Moreover, the question of what function the subgenual organ serves in the case of *Atta cephalotes* cannot be answered unequivocally. It emerges from the research of Markl that it also contributes to the perception of ground vibration, but it has remained unknown for some time what purpose this contribution serves for the ant. Among the non-stridulating ants there are those whose ancestors regressed the stridulator which they had originally possessed; the Dolichoderines and the Formicines belong to these. The

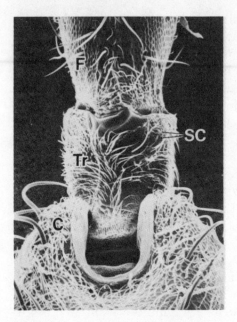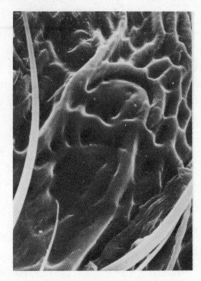

Fig. 3.7   Scanning electronmicrograph of a dorsal section from the leg of a major worker from the leaf-cutter *Atta cephalotes* species (enlarged 360 x). Note the sensilla campaniformia of one of the three groups of sensilla campaniformia on the trochanter-femur joint. C. coxa; F. femur; SC. sensilla campaniformia; Tr. trochanter (Photo – W. Gnatzy).

Fig. 3.8   The same sensilla campaniformia seen in Fig. 3.7 (enlarged 3600 x) (Photo – W. Gnatzy).

regression of the stridulation organ may possibly be connected with the fact that ancestors of these ants relinquished their nests in the solid substratum. However, after many species had changed again to constructing their nests in the substratum, the vibration signal served a useful purpose, had they perhaps not reacquired their stridulating ability, but had developed other vibration signals (Markl and Fuchs, 1972). We have known for a long time that some species, such as the carpenter ants (*Camponotus*), who build their nests in wood, emit a tapping signal when disturbed (e.g. Wasmann, 1893; Gounelle, 1900). A precise analysis of this very rapid tapping of *Camponotus* revealed that the animals first strike their closed mandibles against the ground, lifting up the abdomen and then striking it downwards whilst the head is raised. The reaction of the species mates to this signal — the 'seesawing' of the carpenter ants can be repeated many times — is very variable and depends on the state of excitation of the receiver, as emerged from research undertaken in the natural environment with artificial tapping signals. Slightly aroused animals were stunned by the tapping stimulus, became motionless and therefore invisible to their enemies, who by and large, are only able to localize moving objects. More strongly aroused animals on the other hand were roused to approach the source of the vibration. Apart from this, the attacking threshold of these strongly aroused animals towards visually localized moving objects was reduced. Markl and Fuchs, who conducted this research, concluded from it that the tapping of carpenter ants outside the nest had the function of a theft and danger alarm, in that it increased the effect of other stimuli causing an attack response.

3.4 Sensory organs for light

Ants have as a rule two different types of eyes, two complex, or facetted eyes, and three frontal eyes, or ocelli (Fig. 3.9). Disregarding exceptions, this is true for the workers as well as the reproductive ants. The larvae on the other hand are without eyes; it was not observed for a long time, however, that these maggot-like structures show a reaction to light.

The complex eyes consist of a large number of individual facets (Fig. 3.9). Each of these outwardly visible facets corresponds to an 'ommatidium', which represents a complete eye; it consists among ants of a lens, a crystalline cone and a total of nine retinule cells. Figure 3.10 shows an ommatidium from the complex eye of the wood ant *Formica polyctena,* from the research of Menzel (1972). The external portion of the ommatidium, the lens, is composed of 20-25 layers. The lens is curved round to the crystalline cone, and it can be stained with methyl blue much more intensively in this area than further towards the exterior. This indicates that the lens has a different material composition in the portion curved round towards the crystalline cone. The crystalline cone is surrounded by four cells, which jointly enclose it. The retinule cells, which are situated beneath the lens apparatus of the ommatidium, are radial and arranged round a central axis. Two of these nine retinule cells are particularly narrow (2.1), and one of them rather shorter than the others (9; Fig. 3.11). A structural peculiarity found

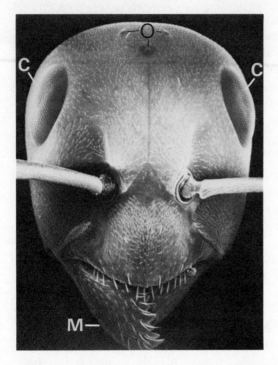

Fig. 3.9   Head of a wood ant *Formica polyctena.* C. complex eye; M. mandible; O. ocelli (Photo — W. Gnatzy).

among bees (Sommer and Wehner, 1975) as well as among the Australian ant *Myrmecia gulosa,* consists of the retinule cells being twisted rather than attenuated. The direction of the retinule cells' twist on one half of the ommatidia is clockwise, on the other half anticlockwise. This rotation of the retinule cells around their central axis amounts to approximately 1° per thousandth of a millimetre. On the inner side, which is turned towards the central axis, the retinule cells carry dense rows of tubular protrusions, in whose membranes the visual pigment is stored. They constitute the light sensitive portion of the receptor cells (Fig. 3.11). The rows of tubular protrusions on all nine retinule cells are fused and together form the 'rhabdome'. This rhabdome has a higher optical density than the surrounding cell bodies. The result of this is that light focussed on the tip of the rhabdome from the lens system is directed downwards, as in a technical lighting circuit. In this way the light is absorbed by the molecules of the optical pigment, which leads by means of a chain of biochemical processes to

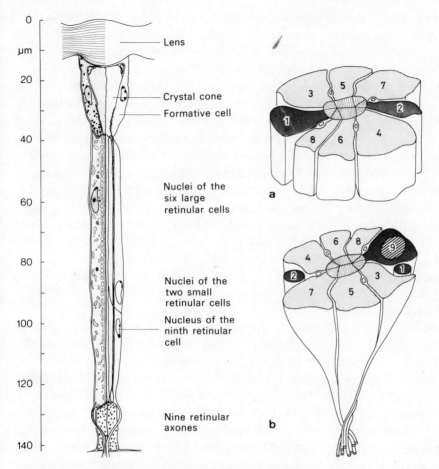

Fig. 3.10   Longitudinal section through an ommatidium of a wood ant *Formica polyctena* worker (after Menzel, 1972, adapted).

Fig. 3.11   Representation of the upper (a) and lower (b) portions of a honey bee's nine optic cells. Eight optic cells are long (1-8); (9) is shorter and only occurs in the lower portion of the ommatidium. The whole collection of optic cells is turned on its longitudinal axis. The centrally positioned rhabdome is shown with the tubular protrusions of the receptor cells; the receptor cells taper off in the lower portion into nerve fibres and are led to the optic ganglia (after Wehner, 1975b).

the stimulus of the sensory cells and to the transmission of a corresponding signal through the nerve fibres leading from it (Wehner, 1975b).

For insects, the perceived environment is composed of numerous image points, corresponding to the number of ommatidia. The greater the number of these image points in relation to ommatidia, the more differentiated the image appears. By comparison, the complex eyes of ants have few ommatidia. Whilst bees' eyes are composed of about 5500 ommatidia, ants with even very well-developed eyes, such as *Cataglyphis bicolor,* do not have more than 600-1300 ommatidia per complex eye (Menzel and Wehner, 1970). At the same time, the complex eyes of the wood ant *Formica polyctena* consist of 750 ommatidia (Menzel, 1972) and the complex eyes of *Myrmica* and some *Lasius* species of between 100-200 ommatidia. Among the South American army ants, *Eciton burchelli,* the 'complex eyes' are reduced to a single 'facet', and among other army ants of the genera *Eciton* and *Dorylus* there is nothing left of the eye. The optical centre of the brain is still developed here during the larval phase; it degenerates, however, in the period between pupation and full adulthood. This is the case only for the females and the workers; the males of these species have very well-developed eyes (Werringloer, 1932).

As far as the functioning of the complex eyes of ants is concerned, it was previously considered improbable that ants were able to distinguish different colours (*see* for example, Sudd, 1967b). This is no longer generally maintained nowadays, however, since the ability to see colour has been demonstrated in the meantime, with the wood ant *Formica polyctena* as well as the desert ant *Cataglyphis bicolor.*

The colour vision of the wood ant *Formica polyctena* was proved by Kiepenheuer (1968). Firstly, Kiepenheuer trained the wood ants to go to a light source, and to carry pupae to it. He later offered them two light sources instead of the one (ultraviolet and yellow/green), at an angle of 90°. The ants then went towards the 'central point of the light', that is on a course between the two light sources, favouring the source which seemed to them to be the brighter. If the intensities of the two light sources were now changed round, so that the ants held a course which exactly bisected the angle between the two lights, then one could be certain that both light sources were seen by the ants as 'equal in stimulus'. With these light sources determined by the ants' light sensitivity, a further group of ants were then trained, so that they had to maintain a course 90° from the yellow light, or 90° from the ultraviolet light in order to reach the nest. In the crucial experiment one lamp only was switched on finally, and the entrance to the nest blocked. If the ultraviolet lamp only was on, then the ants tended to go to the right of the light source; if on the other hand the yellow/green lamp only was on, in the same place where previously the ultraviolet lamp had been, then the ants went in a direction to the left of the light source. The ants showed with these varied behavioural responses to these light sources of different colours, which were clearly, however, perceived as being equally bright, a clear ability to distinguish the different colours.

The desert ant *Cataglyphis bicolor* living in North Africa and the Near East is, with its complex eyes and almost completely optical orientation, a particularly useful experimental animal for optical orientation experiments (Wehner, 1968). Wehner and Toggweiler (1972) demonstrated, with the particularly easily trained workers of this species, that they also were able to distinguish different colours. In contrast to the wood ant, however, *Cataglyphis* did not move in the direction of the 'central point of light', but always decided upon one of the two lights, when offered two light sources instead of one. When these two light sources were sufficiently equal in brightness to be perceived as 'equal in stimulus' by the *Cataglyphis* workers, it was noted that the workers moved towards both lamps in statistically equal numbers. Wehner and Toggweiler then equalized the intensity of two different coloured light sources to such an extent that they were approached by the ants in equal quantities. Later, in separate series of experiments, the ants were trained to move towards the light of one or other wavelength. These experiments were successfully completed by using three pairs of lights of different wavelengths; they demonstrated that *Cataglyphis* workers were able to distinguish colours. The upper limit of the colour vision lay between 630 and 650 nm, and thus below that of humans. This means that *Cataglyphis*, like the honey bee, is no longer able to perceive red light. Wehner and Toggweiler were able to show apart from this that the ants were particularly sensitive within three spectral regions, and that they were able to distinguish particularly accurately between closely related colours within two spectral ranges. This indicates that ants, like honey bees, possess three types of light sensitive cells, which react to specific colours.

Mazokhin-Porshnyakov and Trenn (1972), in their electrophysiological experiments with *Lasius niger,* were able to find indications that three different types of retinule cells are present here, of which two are concerned with various wavelengths within the area where the light is visible, and one reacts to ultraviolet light. This would correspond to the ratios found with honey bees. Indeed, until this time only two types of

receptor cells had been demonstrated with any certainty in the case of
ants, this having been done by Roth and Menzel (1972) with the wood
ant *Formica polyctena,* and by Martinoya *et al.,* (1975) with the
leaf-cutter ant *Atta sexdens rubropilosa.* According to these
electrophysiological recordings, one type of retinule cell reacts
specifically to light at 361 nm wavelength (ultraviolet) and another
type to light with a wavelength of 495-531 (yellow/green).

Which of the nine retinule cells belonged to which of the two types
of sensory cells was determined by the fact that the retinule cells
affected by light which for them was operative 'adapted' in typical
fashion. In this adaptation, small bodies of pigment in the retinule cells
make their way in the direction of the rhabdome. According to Menzel
(1971) and Menzel and Knaut (1973), it was possible to show with the
electron microscope that under the influence of the yellow/green light
the pigment bodies from the six large retinule cells moved in the
direction of the rhabdome, whilst influenced by UV light on the other
hand, the pigment bodies from the two small retinule cells moved
towards it. The spectral sensitivity of the short cells (9; Fig. 3.11), is
not yet known (Ribi, 1975).

The ability of ants, bees and many other insects to perceive the
vibration direction of linear polarized light is linked to the structure of
the complex eye. Light radiating from a source can be described as
vibrations from electric and magnetic fields, which are perpendicular to
each other. The known effects of light upon a photographic film or the
eye are only those caused by the electrical field, which is therefore also
described as the light vector. In natural light, e.g. in sunlight, there is no
tendency towards homogeneity of vibration. When sunlight is filtered
through the atmosphere, however, the electromagnetic waves show a
very definite evening out of the vibrations; it becomes 'polarized'. The
plane of vibration of the sunlight that reaches us by being filtered
through the atmosphere is determined by the position of the sun.
Organisms which are able to recognize the polarization direction of light
can determine the position of the sun by observation of polarization.
This does not amount to much if the sun is visible. If, however, the sun
is hidden behind cloud, then this ability proves to be an advantage to
many insects and other Arthropods. They are thus able to know the
position of the sun by the polarization direction of light filtering
through a piece of open sky, and are furthermore able to make use of
the sun as an aid to navigation.

Without special instruments, man lacks this capability. There are, it is true, many people who are able to see the 'Haidinger's clusters' in a blue sky, the clusters appearing yellow and darker, and perpendicular to the direction of polarization of the particular point in the sky. This effect comes about due to the fact that the yellow pigment in the 'fovea centralis' of our eye can act as a polarization filter (Wehner, 1975b). If, however, we hold in front of the eyes a polarization filter which only allows through light of a certain vibration direction, then we also are able to recognize the appropriate vibration direction at each point in the open sky. One is then able to see that daylight filtered through an electric vector always swings in a perpendicular direction to the plane formed from the viewpoint of the observer of the sun and the particular spot in the sky (Fig. 3.12). Evidently the Vikings were already acquainted with this principle since they, on their journeys to Iceland, Greenland and North America, determined the position of the sun by using polarization sensitive crystals when there was an overcast sky, and were thus able to maintain their course.

How does the insect eye work as a polarization receiver, however? The polarization sensitive structures of the complex eye are the rhabdomes, dependent on the dense rows of tubular protrusions. They

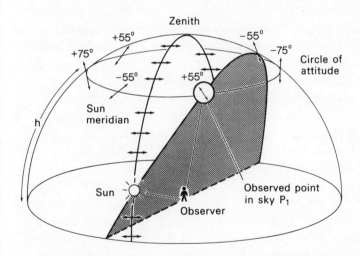

Fig. 3.12   Schematic representation of the polarization of daylight. The vibration direction of linear polarized light is perpendicular to each circumference leading to the sun, as has been shown for the sun meridian and the parallel leading to the point in the sky P. Further points on the parallel of altitude show the vibration directions of the polarized light, which have been drawn in (after Wehner, 1975b, adapted).

ensure that the light running parallel to the axis of the tubules is absorbed almost twice as strongly as the light perpendicular to it. In this way also, one vibration direction of light is more strongly absorbed than all the others, which is described as 'dichroism' (Menzer and Stockhammer, 1951; Täuber, 1974).

The presence of polarization sensitive structures alone in the structure of the complex eye is certainly not sufficient to be able to determine the polarization direction. More receptors are required, with variously oriented analysers. It has been demonstrated with ants that they possess at least two different types of receptors, one for ultra-violet, and one for yellow/green light. Duelli and Wehner (1973) eliminated the polarization perception of the desert ant *Cataglyphis bicolor* which orient themselves in their desert environment almost exclusively by the position of the sun, by making the animals move under a depolarization filter. The ants were then able to maintain their supposed course according to the position of the sun only very inaccurately. If the ants were made to move under filters which filtered out different spectral regions, then it was shown that the orientation capability collapsed when the UV ranges of light were filtered out. This demonstrates that the determination of the polarization direction of daylight occurs by the use of the UV receptors. The use of the UV ranges for polarization measurement is significant for the ants, inasmuch as it is precisely this spectral region which is least susceptible to atmospheric disturbances.

It emerges from the research of Duelli and Wehner that the two receptor types of *Cataglyphis* differ from each other, in that the yellow/green light receptors are insensitive to polarization, whilst the UV receptors on the other hand are polarization sensitive. The calculations of Bernard and Wehner (Wehner, 1975b) have now shown that the UV sensitivity of the long UV receptors is completely lost through the turning of the rhabdome, whilst that of the short cells, as with honey bees, which are also UV sensitive, is maintained. In this way the long UV receptors in bees can be polarization insensitive, and the short UV receptors, polarization sensitive. In order to know the polarization direction, however, two polarization sensitive receptors have their tubular protrusions presented with varying directions. Since, however, there are two prevalent types of rhabdome with tubular protrusions facing in different directions, two related ommatidia, representing both types, are required for the analysis of the vibration direction of the polarized light. They both have polarization receivers facing different directions in the short UV receptors, whilst the long UV receptors serve as polarization insensitive receptors.

We know very little about the function of the three ocelli. Some authors have assumed that the ocelli serve long-distance vision, since these are particularly well developed in insects which are good flyers. Alternatively some authors maintain that they are especially suitable for close vision, and still others assert that they are required for distance localization. Homann (1924) obtained various physical measurements of the ocelli of the wood ant *Formica rufa,* such as the radius of curvature, the focal distance and the aperture of the lens, the distance of the retina from the frontal lens surface etc., and came to the conclusion that the ocelli of wood ants are totally unsuited to image vision. On the other hand, they were apparently relatively luminous, and as useful for distinguishing differences in brightness as for the perception of light direction. Müller (1931) lacquered together the ocelli of female carpenter ants of the species *Camponotus ligniperda,* and then observed that these animals did not shun the light as they usually did and seek shaded spots. On the basis of this experiment he took the ocelli to be exclusively auxiliary phototaxis organs. This small amount of research, however, has not definitively clarified the function of the ocelli.

4 **Orientation among ants**

This chapter is not concerned with orientation in general, but with the particular case of orientation in space. By orientation in space, one understands 'the ability of mobile organisms to alter or maintain their orientation and position in the environment by their own means of control' (Jander, 1970). It follows from this definition that orientation in space usually involves a mobility factor. The movement sequences are regulated, and as a rule triggered, by sources of stimulus outside the organism, as well as being controlled by external stimuli (Görner, 1973). According to the nature of these external stimuli, one distinguishes between optical, gravitational, chemical orientation etc. We have already dealt with the capabilities of ants regarding optical and gravitational orientation, since its investigation is closely linked with the structure of specific sensory organs and with physiological behavioural methods of investigation. Different ant species differ in certain aspects of their orientation quite substantially, such as, for example, the blind army ants and the desert ant *Cataglyphis,* which orients largely optically. On the other hand, most ants probably have several methods at their disposal, which are interchangeable in different circumstances.

4.1 Optical orientation

Ants are able to orient themselves optically in several ways. They can make use of striking objects in their environment as orientation aids, as do the swarming carpenter ants and the sexually reproductive animals of many other species, who direct their flight in the direction of particularly striking landmarks. Ants are also able to use light sources for orientation – this means the sun and the moon in the natural environment – and either move towards or away from the light source (positive or negative phototaxis), or they can maintain a course at a specific angle to the source, which is described as menotaxis. Finally, some ants are able to perceive the vibration direction of polarized light and use it as an aid to orientation.

The ability of ants to orient towards a light source was first demonstrated by Lubbock (1894), who experimented with *Lasius niger.* Lubbock showed in the first series of experiments that the ants did not regulate their movements according to the signs which he had provided for them. They made as little use of the upright pencils as they did of the avenue of small pieces of wood between which he allowed them to move. For a further series of experiments, Lubbock built a turntable, consisting of three concentrically arranged rings which could be turned. He erected this turntable in such a way that the ants were obliged to cross it in order to reach their larvae. If he turned one of the tables, on which just one ant was placed, through 180°, then the animal also turned and continued in the original direction. This reaction from the ants even occurred when the position of the other tables was changed round in addition to the first, so that orientation by odour trails was excluded. The reaction of the ants was different to the turning of the evenly made table if the light, consisting of two candles near the nest, was covered in the laboratory. To this end, Lubbock inverted a box over the turntable. The box contained an entrance and an exit for the ants and an observation aperture for the experimenter. Underneath this box, only 11 out of 30 ants corrected their course after the table had been turned, whilst 19 animals continued in a new direction. If the table was only turned through 90° under the box, then none of the ants corrected their course.

These experiments thus demonstrated that light is of importance for the orientation of ants. Lubbock then removed the box from the turntable, and placed the light on the opposite side of the room from it, at the same time turning an ant through 180°. The result was unequivocal. None of the ants used in the experiment corrected their course, furthermore all the animals continued in the direction resulting from the turn. Finally Lubbock reversed the positions of the candles, but without turning the table, and found that five out of seven ants immediately turned round and continued in the opposite direction.

The fact that ants also make use of light sources in the natural environment was shown by Santschi (1911) in his famous mirror experiment. Using a large screen he shaded workers of the harvester ant species *Messor barbarus* on their return route to the nest, and reflected the image of the sun in the opposite direction using a large mirror. At this, the animals turned round immediately and continued in a direction at an angle of 180° to their previous course. As soon as Santschi removed the mirror the ants turned round again and continued on their old course. With this mirror experiment Santschi was able to lead the ants in any direction he desired.

These experiments by Santschi demonstrated that, under natural conditions, ants are able to orient themselves according to the position of the sun, by maintaining a course at a specific angle to the sun ('light compass orientation'). The changing of the sun's position, which would of necessity lead to the taking of wrong directions with this compass orientation, is, according to Santschi, included in the ants' calculations. They do not adhere absolutely to a specific angle, but alter it according to the position of the sun.

This ability of ants to alter their angle to the sun in accordance with its position during sun compass orientation could not, however, be substantiated by Brun (1914). He enclosed ants (*Lasius niger*) in the natural habitat inside a container and released them again after a known length of time. After this time, the ants altered their original course according to the extent to which the sun's position had changed in the meantime. This emerges, for example, from the following record of Brun (1914): '10 August 1913, 1500 h. After two days of stormy rain. Short sunny periods. A *Lasius* moves during a short sunny spell 7 cm to the left of N and then turns off at a right angle towards the sand area; it walks almost in a straight line, exactly opposite the sun. As it approaches X, the sun is hidden behind a bank of cloud; – the animal makes as if to turn round: at this point I invert a cylindrical tin tube over the ant and trap it inside. It is exactly 3 p.m. During the next two hours it rains almost continually; shortly before 5 p.m. the sun breaks through the cloud. At exactly 5 p.m. I remove the tin cylinder. The *Lasius* sits in the centre of the circle marked by the tube in the sand, its head towards the west. It turns round slowly, placing itself so that its head is facing exactly away from the sun, and walks slowly and almost in a straight line in this direction towards the eastern end of the sand area, arriving approximately 30 cm to the right of this owing to the time lapse. The ant continues a further 7 cm along the bordering stone – exactly as far, in other words, as she had moved to the left at the beginning of the outward journey – it then turns round suddenly and corrects the complete time lapse to the left towards N. (It moreover makes an attempt to turn inwards between the two stones, obviously looking for the nest entrance a stone too soon.) The angle between the primary courses of the outward and the return journeys amounts to 30°

to the right, thus exactly the same as the number of degrees which the sun's arc has travelled across the sky during the two hours when the ant was trapped.' (Fig. 4.1)

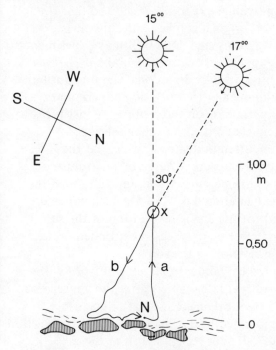

Fig. 4.1   Representation of the sun oriented course of a *Lasius niger* worker according to the research of Brun, 1914. (a) course of the ant before enclosure; (b) course of the ant after enclosure; N. nest; X. place where the ant was enclosed in a lightproof container. (*See* text for details) (after Brun, 1914, adapted).

This experiment indicates a light compass orientation of ants, without, however, confirming the ability of ants to allow for the changing position of the sun. Brun came to different conclusions on the other hand with the slave-making ant *Formica sanguinea;* the workers maintained their original course after being shut in. Brun concluded from these experiments, however, not that ants have the ability to allow for the sun's changing position, but assumes that the *Formica sanguinea* chiefly orient themselves by landmarks on individual journeys. This interpretation of Brun has not been experimentally verified, however.

It has been demonstrated in recent years that many insects, such as bees and beetles, in fact do allow for the changing position of the sun in their sun compass orientation. Jander (1957) subsequently repeated

Brun's experiments, first by determining the course of an individual *Lasius niger* worker, then placing a lightproof cover over the experimental animal and then allowing it to continue after the two hour isolation. The results of 125 individual experiments showed that the ants generally continued on their original courses, and thus clearly behaved differently from the way they had done in Brun's experiments.

Jander also repeated these experiments with the wood ant *Formica rufa,* and applied three different methods: directional training during pupae transportation, flight training and transposition experiments. All three methods were used in experiments on the roof of the institute, and it was concluded from them that ants orient themselves by landmarks. All experiments concurred in demonstrating that the animals' courses before and after enclosure differed by an average of only 2° to the right, and thus remained almost the same, although the sun's position in the meantime had altered by 35°. One can conclude from this that the ants allowed for the changing position of the sun whilst inside the container, and after being released continued on the appropriately adjusted angle in relation to the sun.

In two experiments conducted by Jander in March and April with wood ant workers which had not yet left the nest after the winter rest period, the ants deviated from their original course after enclosure to an extent which on average corresponded to the change of the sun's position during the time of the enclosure. Jander concluded from these experiments that the ability of ants to move according to the compass despite the changing position of the sun, had to be individually learned. It is thus possible to clarify the disagreement between the researches of Brun and Jander with *Lasius niger.* Although Brun conducted his experiments in August, it is possible he may have been working with newly hatched and thus inexperienced workers.

The moon as well as the sun evidently plays some part in the light compass orientation of ants. This emerges from observations on *Monomorium salomonis* (Santschi, 1923) and from experiments with the wood ant *Formica rufa* (Jander, 1957). Jander discovered that among the wood ants the workers also gather together their brood and transport it into the nest during a full moon, Jander having previously scattered the brood throughout an arena. When the moon was hidden from the ants and reflected in another direction, then the brood-carrying ants adjusted their course by the appropriate amount.

Jander investigated the mechanism of light compass orientation with *Formica rufa.* He made a distinction here between 'ground orientation' (negative and positive phototaxis) and 'derived orientation' (menotaxis). The characteristic of ground orientation is that it is innate, and generally used only for adjustment possibilities: in the case of positive phototaxis adjustment towards the light, in the case of negative phototaxis adjustment away from the light. With derived orientation it is different. With menotaxis many courses are open to the animals in practice, courses at their disposal depending on the source of the stimulus, and which of the possible courses is adopted is the result of a learning process.

Jander found both types of orientation with *Formica rufa*. Newly hatched ants and those leaving the nest for the first time orient themselves phototactically. One can see by this that the young ants and also the queens immediately seek out the darkness, if one uncovers the nest (negative phototaxis), and that the ants on their first journeys outside the nest move in the direction of the sun (positive phototaxis). On their return journey, these animals move away from the light, thus finding their way back into the nest, in whose vicinity — and probably not only just here — other orientation mechanisms are added. The phototactic course of ants does not, as a rule, run in a straight line; more often they oscillate in their course from one side to the other of a mean line. An ant which exactly maintains its course on the mean line is situated on the *mean site*; the course between the mean site of the animal and the centre of the object is described as the *fixed course*. The angle α between the direction of movement and the fixed course is $0°$ with positive phototaxis and $180°$ with negative phototaxis. The point in the eye which is stimulated in the mean site is called the fixation point. The further an animal moves away from the mean site, the stronger will be the tendency of the ant to turn.

Jander offers various suggestions how the central nervous system controls the ability of ants to direct their movements away from, or towards a light source. An integration centre compares the intensities with which the individual ommatidia of the complex eye are stimulated, and acts upon a position centre, in which these values are compared with those of other sensory organs, e.g. the gravity sensitive organs. For its part, the position centre influences the coordination mechanism of the legs, leading finally to the appropriate turning movements. The physiological distinction between positive and negative phototaxis remains unclarified. Apart from this, it is still not known in what way a long unknown agent ensures that ants sometimes orient by positive phototaxis and at other times by negative phototaxis.

Menotactic orientation is essentially more plastic than phototaxis. It allows the ant to maintain a course at any desired angle to a source of light, the choice of angles being dependent on the learning process of previous journeys. It was assumed in bygone years that ants returning to the nest followed the exact course of their outward journey. According to Piéron (1904), the ants store adjusting movements and turns collected during the outward journey in a reversible 'kinaesthetic memory', and reiterate them in reverse sequence on the return journey, thus eliminating a measure of the fatigue resulting from the distance. According to present-day understanding, the central point of kinaesthetic orientation is that it proceeds without 'the direction of external stimuli' (Görner, 1973), and that 'spatial information is drawn from the signals produced by the organism itself, signals which are necessarily linked with spatial conditions and changes under normal conditions' (Mittelstaedt and Mittelstaedt, 1973). Orientation of this kind according to self-accumulated data is defined by Mittelstaedt and Mittelstaedt as *idiothetic orientation,* distinguishing here between directional and deviational idiothetic orientation. In cases where the spatial information about the relative position of the animal is drawn from a spatially arranged physical factor, one speaks correspondingly of *allothetic orientation* (Mittelstaedt and Mittelstaedt, 1973).

The wood ants investigated by Jander (1957) were, in fact, able to learn on both outward and return journeys and thus ascertain the direct course. He was able to show that this direct course to the nest was not so easily found after one journey as after repeated journeys. The discrepancy between the actual course run and the path of the direct course, described as the 'residual discrepancy', is a transient phenomenon and probably does not make very much difference under normal conditions. According to Jander, the wood ants are not able to learn the path of a course by idiothetic data alone. This does not altogether exclude, however, the possibility that with other research methods, such as perhaps those applied by Mittelstaedt-Burger (1972) to other insects ('compromise adjustment'), an idiothetic component in the orientation of ants will nevertheless be possible to prove (Görner, 1973).

Whilst in phototaxis the fixation point in the eye of the insect is inborn and there are only two such points, one for positive and one for negative phototaxis, there are in menotaxis many possible fixation points, and the question still remains unclarified how these points now used in menotaxis came into being. According to the compensation theory of menotaxis, put forward by Jander, the phototactic ground orientation of experienced ants which orient themselves by menotaxis is not in fact obscured, but continues to function. If the fixation point lies in the frontal eye region, this indicates positive phototaxis, if it lies in the rear part of the eye, then negative phototaxis is in operation. In order to deal with the discrepancy between the ground orientation course and the current menotactic course, there is, according to Jander's theory, a command centre, which acts upon the location centre together with the integration centre and other agents producing a 'turn command', by which the fixation points of both eyes are shifted. On the return journey to the nest, the disposition of the ants changes from being centred on the outward direction to being centred on the nest, leading to a change from positive to negative phototaxis, to the shifting of the fixation point front to the rear part of the eye, and thus to the turning of the ant through 180° (Fig. 4.2).

In reality the orientation of ants is certainly much more complicated, since apart from further optical orientation aids (polarization patterning of daylight and use of landmarks) there are still more to be added, such as rises and dips in the vicinity of the nest and also chemical trails laid down by the ants.

The ability of ants to orient themselves according to the vibrational direction of polarized daylight, was first demonstrated by Schifferer (after von Frisch, 1950) with *Lasius niger*. He trained the animals to find their feeding place underneath a polarized film. He then turned the film a specific number of degrees and was able to observe that the ants immediately deviated from their previous course by the same amount. Vowles (1954b), Carthy (1951) and Jander (1957) also experimented with polarization film, and were able to show that apart from *Lasius niger* also *Myrmica rubra* (Vowles), *Formica rufa, Tetramorium caespitum* and *Tapinoma erraticum* (Jander) were able to make use of the vibrational direction of polarized light as an aid to orientation.

Orientation by the vibrational direction of polarized light and sun compass orientation are not contradictory under normal conditions; they rather complement each other, since both orientation aids in common adjust according to the changing position of the sun. If then the polarization direction of daylight is placed in such a way as to be antagonistic to the position of the sun, shifting the polarization direction in opposition to the sun's position by, for example 50°, then in the case of the wood ant *Formica rufa,* the ants proceed on a

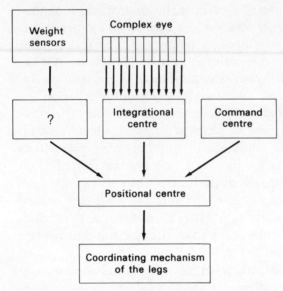

Fig. 4.2   Schematic representation of the effects of various central nervous agents upon the orientation of red ants (after Jander, 1957). (*See* text for details.)

compromise course. If one shifts the polarization direction by 180° against the sun's position, then the wood ants direct themselves without exception by the vibrational direction of the polarized light (Jander, 1957). On this basis, Santschi's mirror experiment is unsuccessful on a number of counts with wood ants and a range of other ants.

In order for ants to be able to orient themselves according to the polarization pattern of the sky, it is not sufficient that the ants are capable of determining light vectors at various points in the sky. In addition to this, they must be able to predict the future position of the sun from their information of the complete polarization pattern. On account of the symmetrical properties of the polarization pattern, however, one point only in the sky provides ambiguous information, as emerges from Fig. 3.12. Thus it is true that an ant is able to determine the sun meridian by looking at one section of the sky in the zenith, the meridian running perpendicular to the polarization direction in the zenith, but it is not able to distinguish between the two halves of the meridian. The ant will as a result count two directions deviating by 180°

as equal, as is shown in experiments with ants and honey bees, which are only able to see a small section of sky (Wehner, 1975b).

In fact, however, under normal conditions there are generally more and larger areas of sky available to ants. The question now is whether they calculate their course from a chosen area of sky with a proportion of their ommatidia, or whether they sight a specific area of sky with all their ommatidia. Duelli (1974, 1975) was able to come to a decision between these alternatives, both by partially lacquering together the eyes and by reducing the 'sky vision' of the animals by placing a rotating 'planetarium' over the ants, in order to limit their field of vision. It was shown in these experiments that nearly all areas of sky are equally useful for the determination of a navigation course, but that only specific areas of the eye are designed for these measurements.

This also agrees with research carried out on the eyes of bees, during which Frisch (1965) found that a specific area of the eye, with a diameter of 11-15°, suffices for determining a polarization course. Zolotov and Frantsevich (1973) specify that a minimum of 25-50 ommatidia is required. The smallest area of eye that will statistically 150-200 ommatidia are required for full orientation capacity. In the desert ant *Cataglyphis bicolor* an area with a diameter of 20° with 30-60 ommatidia are required. The smallest area of eye that will statistically still permit orientation by polarization pattern, comprises 10° with 7-19 ommatidia respectively (Duelli, 1975).

As soon as one shields the polarization sensitive area of the eye of a desert ant from daylight, the ants correct the position of their heads in such a way that this polarization sensitive area of the eye is once again directed towards the open sky. A shielding of, for example, 15°, brings about a raising of the ant's head by exactly 15°, so that the polarization sensitive area of the eye is once again able to perceive the amount by which the light is being shielded (Duelli, 1974, 1975).

Under natural conditions, specific ommatidia form images of specific areas of sky. When, however, the ants are compelled in the experiment to raise their heads, then they must, for the correct optical orientation, obtain information about how far they have raised their heads. This information is in part provided by the bristle areas of the neck joint, which as proprioreceptors measure the position of the head in relation to the rest of the body. Information from the mechanical receptors also contributes to gauging the altitude line system to the spatial constant (Wehner, 1975a,b). Apart from the neck joint the legs are also concerned with this measurement, not only adding to the compensatory capacity of the neck joint by tilting the thorax, but also by taking over the larger part of the task. Compensation by the forelegs is also controlled by proprioceptive bristle areas. The distribution of compensation into two different mechanisms occurs, according to Wehner, for reasons to do with the technicalities of measurement, namely that if the neck joint alone took over the task of altering the angle, the sensory bristles concerned with this would be bent back so far that their dynamic measuring range would be impaired, and they would no longer be able to signal the angle change sufficiently accurately. Through the support of the forelegs, the neck organ is constantly maintained in the performance range most favourable to this measurement.

Ants are able to orient themselves by landmarks as well as by the condition of the sky, as emerges from the very early research of Wasmann (1901). Jander (1957), in his research with the wood ant *Formica rufa,* placed landmarks in competition with sky conditions. He firstly trained the *rufa* workers in the institute garden to find their nest by a particular course. He then set up the experimental area on the roof of the institute, and trained the ants to look for the nest along a course which deviated from that of their initial training by 90°, according to data from the sky. If the ants were then brought back into the institute garden and the experimental area set up again as it had been for the initial training, the ants were then presented with a conflict. The landmarks (initial training) indicated a different course from that indicated by sky data (second training). The ants responded to this situation in such a way as to adopt a course which compromised between those of the two training courses, thus showing that they utilize landmarks as well as sky data for orientation. If there are no sky data at the disposal of the animals, as sometimes occurs with a heavily overcast sky, then the wood ants direct their course exclusively by the landmarks. The orientation of ants by landmarks presupposes that they are able to compare various landscape images and to perceive the differences. For this purpose, there must be some agency in the central nervous system of ants that refers the various complex landscape images to some specific pattern and then decides upon the course to be followed. According to Jander (1963) such an agency is described as a *detector.*

We know that bees and bumble bees possess various detectors. A 'pattern detector' dominates during the feeding flight, responding to the spatial location, distribution and abundance of light/dark boundaries. During the flight of bees and bumble bees to the nest, a so-called 'dark-centre detector' predominates, whose optimal response is towards dark objects of a specific size.

Voss (1967) found various detectors among the wood ants (*Formica rufa, Formica polyctena* and *Formica nigricans*): a 'darkness detector', differing from the dark-centre detector of bees and bumble bees in that it does not respond maximally to any specific size of dark object, but reacts with equal intensity to all dark objects over a certain size. By way of a second detector, Voss also found a 'brightness detector', corresponding to the darkness detector. The third detector found by Voss was a 'vertical border detector', which responds to contrasting black and white lines running vertically. These three detectors appear to be inborn in ants. Newly hatched ants spontaneously favoured white and completely black surfaces over patterned surfaces on their path, and favoured vertically running stripes over horizontal. It is possible through training to bring ants to favour, against their spontaneous inclination, bi-coloured surfaces over completely black or completely white surfaces. The vertical border detector, on the other hand, cannot be affected by training; one cannot, therefore, train ants to favour horizontally running stripes over vertically running stripes. Voss has described a fourth detector as a 'segmentation detector', which responds only to the degree of segmentation but which is not sensitive to changes of form in the contrasting pattern.

Wehner (1968) investigated the connection between orientation by landmarks and by the position of the sun, using the desert ant *Cataglyphis bicolor*. Wehner captured individual workers outside the nest in their natural environment in Israel and then released them at various distances 180° from their original course on the other side of the opening to the nest. In this way, landmarks and the position of the sun were antagonistic: if the ants continued to follow their sun course, they would move further away from the nest. These evidently purely optically oriented ants, however, found their way back to the nest, merely by orienting themselves according to the landmarks. The results showed that the ants found their way back to the nest the more easily, the closer to the nest they were released. This indicates that in the vicinity of the nest the ants orient themselves contrary to the previously adopted sun course, finding their way back to the nest on the basis of landmarks. In areas away from the nest, however, at distances of around eight or more metres from the nest, the ants are evidently unable to make out any more landmarks and hold to their sun course, which conveys them further away from the nest. In the absence of all optical aids to orientation – as Delye (1974) observed during a total eclipse of the sun – the *Cataglyphis* workers are utterly incapable of orienting themselves. They remain motionless on the spot, wherever they find themselves.

4.2   Chemical orientation

A large number of ant species lay trails with certain glandular secretions, which can be utilized by themselves and their nest mates as aids to orientation (*see* Chapter 5). If up to now we are still not sure of the sensory organs with which ants perceive this trail substance, nevertheless behavioural research, such as with the army ant *Eciton* (Vowles, 1955) or with *Lasius fuliginosus* (Hangartner, 1967), indicates that they are situated on the antennae and respond to aerial stimuli, in contrast with the termite *Trinervitermes*, which perceives its trails with taste receptors (Tschinkl and Close, 1973). The trail substances of the leaf-cutter ant *Atta texana* are also perceived as odorous substances. This emerges from the experiment of Tumlinson *et al.,* (1972), who put down an artificial trail and made the ants walk along a film of plastic 1 mm above the trail. The experimental animals were able to follow the trail under these conditions, thus showing that they were orienting themselves by the odour field of the trail.

The odour field of an ant trail was investigated in more depth by Wilson and Bossert (1963) and by Bossert and Wilson (1963) with the fire ant *Solenopsis saevissima*. It is three-dimensional, and so constituted that the evaporated trail substance is strongest along the odorized line, and decreases away from its sides (Fig. 4.3). This differential concentration of odour substance in the odour field of the trail is the only factor indicating the course of the trail for the ants. One can either draw various scent probes, compare them with one another and ascertain the course by a 'palpation of the odour area' in such a manner; or one can follow an odour trail with unpaired or unilateral sensory organs, in this case orienting oneself clinotactically. Animals with bilateral sensory organs are able to draw two scent probes simultaneously from separate points in the odour field, compare them with each other and adjust their course in such a way that both sides present an equal odour intensity: as long as this remains so, the animals are sure to continue along the odour trail. This second method of orientation within an odour field, by the simultaneous comparison of two probes, is described as osmotropotaxis.

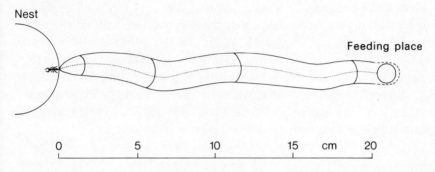

Fig. 4.3 Odour field of the trail of the fire ant *Solenopsis saevissima*, laid by an animal on a glass plate between the feeding place and the nest. The concentration of the odorous substance falls below the behaviour-triggering threshold after about 100 s (after Wilson and Bossert, 1963).

Hangartner has demonstrated that both orientation methods are possible for ants, using the example of *Lasius fuliginosus*. In this experiment, the animals normally oriented themselves tropotactically, by comparing odour intensities measured simultaneously by both feelers, and so adjusting their course that the measured intensities from the two feelers were equal. This emerges from the following experiments:

Hangartner made three groups of ants follow a trail, the first group were left as they were, the second group had one feeler removed and the third group had their feelers crossed over and stuck together, so that the left hand feeler pointed to the right and vice versa. This showed that the untreated animals followed the trail the most quickly, the group of ants which had had one feeler amputated required twice as long and the group with their feelers stuck together required all of five times as long. The time loss of the second group was due to the fact that the animals had been obliged to palpate along the trail with only one feeler, whilst that of the third group was due to the fact that they were deceived by their crossed feelers as to the direction of the deviations and then compensated to the wrong side of the trail, thus only finding their way back to it by a process of trial and error.

It emerges from a further experiment by Hangartner (1967) that ants are able to measure and distinguish between two different odour intensities simultaneously. Two parallel odour trails were laid out a short distance from each other, at first equal in strength, but one of which became continually weaker. At first, the ants followed these trails by walking between them, palpating the odour fields of the two trails with their feelers (Fig. 4.4). As soon as the odour intensity of the weaker trail fell below a certain amount, the animals changed over to the other trail, and followed this on a slightly undulating course (Fig. 4.4), which allowed the conclusion to be drawn from the experiment that they were maintaining a balance between the odour intensities measured by the two feelers.

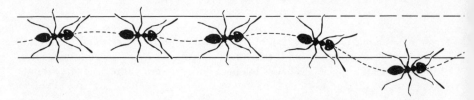

Fig. 4.4   Course of a *Lasius fuliginosus* worker along a double trail, of which one side is weaker in intensity. The concentration of the odorous substance is twice as strong along the continuous line as on the dotted line (after Hangartner, 1967).

If the *fuliginosus* workers are offered a forked trail, the ants only follow both branches of the trail in equal numbers when the odour concentration is equal on both branches. One can determine the ratio of division of the ants from the numbers deciding on the trail substance on the two branches (Hangartner, 1969a). This is only valid in a laboratory environment free of optical stimuli. In the natural environment, this principle will be rendered void by the additional optical stimuli, particularly of the experienced animals. One can see the additional optical orientation from the fact that the course of the chemical trails is generally exactly the same in a number of consecutive years (Gaspar, 1967).

Ants are able to orient themselves optically at the same time as following a chemical trail. This emerges from experiments by Marak and Wolken (1965), who allowed *Solenopsis saevissima* to follow a scent trail, at the same time as illuminating the trail on one side. If they put out this light and lit the trail from the other side, the ants immediately turned round and followed the trail in the opposite direction.

With the wood ant *Formica polyctena,* optical orientation clearly predominates over chemical. This is shown in experiments by Rosengren (1971), who allowed the wood ants in the natural environment to choose between optical and olfactory stimuli. Chemical orientation could, however, play a part under three circumstances (Horstmann, 1976). 1. In an acute shortage of food – which in any case occurs only rarely – the scout workers lay a trail on their return journey, the trail being stronger in odour than the trail of an equal number of workers from a nest where there is sufficient food (Horstmann, 1975b); 2. apart from times of scarcity, wood ants also lay trails from honeydew sources to the nest, which could chiefly serve the young ones among the outside workers as an aid to orientation; 3. finally, the trail of wood ants could also help them to orient themselves on dark nights. Rosengren (1971), observed that such orientation by night was possible.

4.3   Orientation by gravity

The ability of ants to orient themselves by gravity has already been discussed in connection with the gravity receptors. Ants can maintain any desired angle in relation to gravity on a sloping surface and thus utilize the gravity stimulus as an aid to orientation. Such a method or orientation, where a specific angle is maintained in relation to gravity, is described as geomenotaxis. Geotsch (1934) has already demonstrated the existence of geomenotaxis for *Crematogaster scutellaris,* Vowles (1954b) for *Myrmica ruginodis* and *Myrmica rubra* and Markl (1964) for *Formica polyctena.* Young and inexperienced workers, on the other hand, as Jander (1957) showed for *Formica rufa,* move in the same direction as gravity: they are positively geotactic, on the return journey to the nest negatively geotactic. This shows that as with orientation by a light source, a ground orientation – here geotaxis, which is obviously innate – and a derived orientation – here geomenotaxis, which presupposes a learning process – are in operation.

On the basis of research into gravity orientation, Vowles (1954b) concluded that the workers are indeed able to maintain a specific angle in relation to gravity, but that they are not able to distinguish between two symmetrical angles – with reference to the vertical – (α and $360° - \alpha$). Such an inability would, moreover, lead increasingly to misdirections, and would indicate that the ants are not able to use geomenotaxis a great deal. Markl was able to show – in contrast with Vowles – that the

Formica polyctena and *Myrmica rubra* workers investigated by him were able to learn any desired angle in relation to gravity and not to confuse it with any other angle.

In Markl's experiments, the ants were trained to find a food source at a specific angle in relation to gravity. It was shown here that the precision of geomenotactic orientation in ants is limited. The ants deviated as a rule from the correct course, then, however, 2/3 to 4/3rds of them began to move round in circles and usually found their goal quickly. This deviation at the correct point of the search path presupposes that the animals are able to estimate how much of the distance they have already covered. It is still not known by what physiological means this is successful for the ants.

The deviation of the geomenotactically oriented ants adjusts itself, according to Markl's research, and rectifies itself as a rule. The ants deviated upwards from the actual course, when this deviated by 30°, 60°, or alternatively by 330° and 300°. They deviated downwards when the training direction varied by 120°, 150° or alternatively 240° and 210°. Markl was unable to ascertain any deviations in the training courses lying between these figures. It was similar on the return journey: the deviations tended downwards from the theoretical course by 210°, 150°, 240°, 120°, 270° and 90°, upwards from the theoretical course by 80° and 90°.

The cause of this geomenotactic erratic orientation (Markl, 1964) has remained unknown for a long time. We only know which of the possible alternatives is not the case. Markl was thus able to show that the erratic orientation was not the result of an initial error, as occurs temporarily during the acquisition through learning of a photomenotactic course. Nor does it have a mechanical cause, such as that displayed by the ants when moving diagonally to the direction of gravity of a strong turning moment on their high axis. Nor does the erratic orientation have its roots in peculiarities of the gravity sensitive organs.

Although geomenotactic erratic orientation is not reminiscent of a misdirection of optical orientation, there are, nevertheless, certain features in common between gravity and light orientation. In both cases one finds an innate ground orientation (positive and negative geotaxis and phototaxis) and a derived menotaxis, in which a specific angle is maintained in relation to the source of the stimulus, which is thus learned from light or from gravity. Optical and gravity orientation are connected by a location centre in the central nervous system (*see* Fig. 4.2), where an orienting light stimulus can be replaced by a gravity stimulus (Vowles, 1954a; Jander, 1957). The ants are thus able, without having to learn, to transpose an angle in relation to a light source into a corresponding angle in relation to gravity. A similar ability has already been demonstrated in the case of the honey bee, which is able to transpose – by dancing on the perpendicular honeycomb – an angle of direction obtained by light orientation into gravity orientation.

Ants only orient themselves exclusively geotactically, or phototactically in the rarest cases in the natural environment; they generally orient themselves by a combination of the two, and combinations of both or one method with chemical orientation. The observations and experiments of Goetsch (1934) showed a combination of geotactic and chemical orientation: workers of *Crematogaster scutellaris* oriented themselves geotactically when they had to climb a perpendicular wall on the return journey to the nest. If they were offered a fraction of their chemical trail on a piece of paper, they would follow this trail, deviating not more than 40° from their geomenotactic course. If the direction of this course differed by more than 40° from the gravity course of the ants, then they were undecided in their movements. In the case of its differing by more than 75°, the ants followed their geomenotactic course and left the odour trail without paying it any attention.

5 Communication among ants

5.1 The necessities and possibilities of communication among ants

The ability to communicate is one of the essential preconditions for
social behaviour; with their highly developed social life in all its
numerous expressions, ants possess a particularly full repertoire of
possibilities for mutual understanding. Communication signals are
important for raising the alarm when defending nests, as a predator
alarm when being overpowered by predators, tracking substances for
orientation towards a food source, or for recruiting nest mates, and as
sexual bait for finding and recognizing sexual partners. Within the
colony also, the various members, such as larvae, pupae, workers and
queens, are recognized by specific signals and there are even special
chemically recognizable signals for dead nest mates.

We have already discussed two examples of communication among
ants; trapped leaf-cutter ants use the stridulation signal as an emergency
alarm, inducing the nest mates to come to them and dig out the trapped
animals. Among carpenter ants, and probably also among the weaver ant
species *Oecophylla longinoda* (*see* p. 74) tapping signals serve to arouse
the nest mates. In these two cases, it is mechanical signals which serve
for the communication of ants in a colony.

It was formerly assumed that ants communicate chiefly by
mechanical means. This construction came about not only because of
the relatively early discovery of the stridulation organ of ants – it was
still not know what purpose they served – but mainly because it had
been observed on frequent occasions that ants manipulate their feelers
reciprocally with varying intensity and frequency. The existence of a
'feeler language' of ants was concluded from this (e.g. Wasmann, 1899).
Later, after more systematic behavioural research had been conducted
and a series of glands had been discovered, authors increasingly
surmised that the chemical communication system of ants could be
even more significant (e.g. Schneirla, 1952; Goetsch, 1953; Le Masne,
1953 and Stumper, 1953) and they were to be proved right. Shortly

after, the first signalling substance was successfully analysed, giving rise
to an abundance of further work, during which a large number of
chemical signals were discovered and analysed. How much has been
gained from this work can be seen perhaps from Priesner's survey
(1973). We are nevertheless probably only in the initial stages of a line
of investigation whose extent and significance cannot yet be perceived
in any detail.

Despite the great significance of chemical communication for the
social life of ants, the outstanding representatives of chemical
communication are not ants, but butterflies. It was discovered in the
last century that male butterflies can be lured by females of the same
species from distances of several kilometres. Then, at the turn of the
century it was discovered that this attraction is linked with a gland
situated in the rear section of the female body. Liquid substances are
produced, released by these glands and scented by the male. As soon as
the males sense this sexual bait, they fly in the direction of increasing
concentration, thereby finding the females. This connection, however,
between the glands of female butterflies and the attraction of the males
was only discovered relatively recently. The reasons for this are
probably to be found in the fact that we were not able to isolate the
substances produced by the glands, and were thus unable to experiment
with them. Apart from this, we would not have imagined that scent
substances could be effective over such distances. Finally, a more
important reason for their late discovery lay in the fact that these
sexual attraction substances are odourless to humans (Schneider, 1970).

The signalling substances of ants are also odourless to humans to
some degree, at any rate in the concentrations effective among ants
themselves. These signalling substances were earlier called ectohormones
(Bethe, 1932), in order to indicate the similarities between them and
hormones. Hormones act as messenger substances within the organism,
'ectohormones' perform the same task, but between individuals. Of
course, if one conceives of hormones as the products of glands which
secrete inside the body, then the concept 'ectohormone' means a
contradiction in itself. This was pointed out chiefly by Karlson, for
which reason he – together with Lüscher – suggested that signalling
substances between different individuals be described as a sort of
pheromone (Karlson and Lüscher, 1959). According to this designation,
pheromones are substances given out by an animal which either trigger
a specific behavioural response in another animal of the same species, or
produce a physiological development (*see also* Karlson and Butenandt,

1959). According to Brown (1968) and Brown *et al.* (1970), there are
two separate designations for substances acting between organisms of
different species, according to whether the producer or the receiver
profits from the communication. Substances effective between different
species and from which the producer gains an advantage, such as in the
case of warning and defence secretions, are called allomones; substances
by which the receiver profits, such as in the case of the scents of forage
plants as well as those of hosts and predators, are called cairomones. We
shall see that the boundaries between these various substances are
somewhat fluid, and that some pheromones may also act as allomones
or cairomones.

Pheromones are effective in different ways in the receiver individuals,
according to whether they trigger a certain behavioural response or
determine a physiological development. In the former case one refers to
a 'releaser', in the latter to a 'primer'. Pheromones which trigger a
specific behavioural response act upon the smell and taste organs
respectively. From there, information proceeds to the central nervous
system which, for its part, brings about the behavioural adjustments
through effectors. By contrast pheromones acting as 'primers' are, when
possible, taken by the receiver individual by mouth, are reabsorbed by
the stomach and intestinal tract and take their effect via the blood
circulation, to which we have not yet referred.

It is generally the case, however, that chemical, mechanical and also
optical signals act together in communication among ants. For this
reason it would seem more useful, instead of ordering communication
according to the signals used, to proceed from the behaviour and
demonstrate which behavioural forms are triggered by which signals.

5.2 Triggering of alarm behaviour

It was observed some time ago that the sting of a single bee caused an
attack by more bees, and it was surmised because of this, that their
poison triggered the attack behaviour (Huber, 1814; Buttel-Reepen,
1900). Maschwitz (1964a) then demonstrated that in fact other
glandular secretions were responsible for this. Firstly, he squeezed
worker bees with pincers at the entrance to the colony, and observed
that some workers would thereupon take up guard positions or

alternatively fly off and attack objects in the vicinity of the colony. This alarm behaviour was not only released by complete bees, but also by a proportion of those freshly killed bees, perhaps with squashed stinging apparatus, which were laid in front of the flight hole. Maschwitz then further prepared the stinging apparatus and investigated the different component parts for their effectiveness as triggers of alarm behaviour. He discovered through these investigations that it was not the poison gland, but the drip pad of the sting which was effective, and that the mandible gland of the bees also produced a secretion which stimulated the animals to attack. We have since learned through chemical research that the behaviour stimulating substance of the drip pad of the sting is isopentylacetate, while that of the mandible gland is 2-heptanone, isopentylacetate being far more effective in triggering the alarm behaviour than 2-heptanone (Boch *et al.*, 1962; Boch and Schearer, 1965, 1967; Free and Simpson, 1968).

Alarm behaviour in ants is also chemically triggered as a rule. Owing to the large number of species, one finds here not only an abundance of different alarm substances, but also different reactions to the same stimuli. Some species react to the alarm substance with an attack response, others with flight or by playing dead, such as the workers of *Tapinoma melanocephalum,* among which Goetsch (1934) found the first indications of the existence of alarm substances. According to the observations of Maschwitz (1964a), alarm behaviour occurred to a greater or lesser extent in all the species he investigated, always when the alarm substances were presented in the nest, but also in some species such as *Formica* and *Crematogaster* if one presented pheromones on the edge of the trails between the feeding place and the nest or, as with *Formica polyctena* and *Tapinoma erraticum,* if one placed the substance at the feeding place itself. The aroused ants move about under the influence of the alarm substances at greater speed, but in the particular direction of the source of the alarm substance, and approach with threateningly open jaws. Objects treated with the alarm substance were thoroughly felt by the antennae and — as far as possible — dragged away. The intensity of this form of behaviour varies from species to species. It can also occur that individual workers attack each other under the influence of alarm substances, then 'recognize' each other and free themselves from the attack. Maschwitz was only able to observe a flight reaction to alarm substances at the place of feeding, and only with *Lasius niger, Tapinoma erraticum* and *Aphaenogaster testaceo-pilosa.*

Aphaenogaster was fed directly at the entrance to the nest with sugar solution, the two other species at a distance of up to 2 m from the nest. As soon as the alarm substance of their species was presented to them, the animals fled immediately or after a short time in the direction of the nest. Newly arrived ants turned round immediately.

It emerges from experiments by Maschwitz (1964a) with *Myrmica* workers that alarm substances are actually given off by ants in danger. Maschwitz squashed individual animals and thereby determined that the strongly smelling mandible gland substance was given off, which triggered the alarm behaviour. There were, however, differences to be observed among the individual species as to the promptness with which the substance was released. *Myrmica sulcinodis* caused an alarm response in 18 out of 20 cases, *Myrmica ruginodis* and *Myrmica rubra* by contrast in only five out of 20 and *Manica rubida* in four out of 20 cases. In a fight between two species of *Myrmica,* the smell of mandible gland substance was also clearly perceptible. In aroused *Messor* workers, the alarm substance appeared in the form of droplets at the tip of the outstretched sting.

The predator alarm is usually distinguished from this danger alarm, the predator alarm serving to bring reinforcements against large and powerful predatory animals. Maschwitz investigated the predator alarm in the forest ant *Formica polyctena.* He was able to show in laboratory experiments that meal beetles placed in the area of a laboratory colony would attract more ants, and be repeatedly bitten by them, if they had previously been smeared with the forest ant alarm substance. Maschwitz showed that the predator alarm is also effective in the open by loosely binding a large cockroach (*Nauphaeta spec.*) well able to defend itself, and pinning it with a thread 4 cm long leading from the thorax in the middle of a cavity 2 cm deep and 10 cm wide, in such a way that it was able to move freely. A ridge running round the cavity ensured that the cockroach could not be seen by ants moving past it. Maschwitz then counted the ants which went to the cockroach after it had been found and attacked by one ant. Cockroaches which had not yet been found by ants and empty cavities without cockroaches served as controls. This experiment showed that the forest ants finding a predatory animal and spraying it with poison were able to fetch assistance from within the immediate vicinity, up to 10 cm from the ridge of the cavity. In this case, the same substance (formic acid) serving as a chemical weapon is also the alarm substance. This is the case not only with *Formica polyctena,* but also for other *Formica* species, such as *Formica cinerea,*

Formica fusca, Formica rufa (Maschwitz 1964a; Löfqvist, 1976) and similarly for *Camponotus pennsylvanicus* (Ayre and Blum, 1971). At the same time, they use formic acid to raise the alarm to patrolling nest mates, who arrive at the spot in particularly large numbers in a matter of seconds.

There are also some, albeit only a few, ant species which produced no alarm substance in the experiments of Maschwitz (1964a,b). Belonging to this group of species were those with small colonies and few workers (mostly under 100) such as *Ponera coarctaca* and *Myrmecina graminicola.* In the correspondingly small nests of these species, any danger can probably be directly perceived by all the occupants, thus probably rendering a chemical alarm system superfluous.

As far as other species are concerned, however, alarm pheromones are widely distributed and have thus far been demonstrated in six out of 10 of existing subfamilies. Among Ponerines and Dorylines the alarm substances are formed exclusively in the mandible glands, among the Dolichoderines, in the anal gland lying in the rear section of the body. Apart from the mandible gland, Myrmeciines also produce alarm substances in the end of the intestinal tract and in Dufour's gland, whilst the Myrmicines and Formicines produce it in the mandible and Dufour's gland — a few Myrmicines also produce alarm substance in the metathoracal gland — as well as in the poison gland (Fig. 5.1).

The alarm substances serve to a considerable extent in the defence of the colony, as the behavioural experiments have shown. It is possible, in fact, that they originally developed out of the defence substances themselves as a deviation. This is indicated by those examples in which the alarm and defence substances are identical, such as among *Formica* species, which make use of formic acid both for defence and for raising alarm. The alarm substances from Dufour's gland in the Formicines also serve in defence in that, apart from being poisonous themselves, they reinforce the effect of poisonous defence secretions. In the Formicines they are always given off together with the secretions of the poison gland. This emerges from the research of Morgan and Wadhams (1972), who found no muscles in the Formicines which would permit the separate emptying of the Dufour's and poison glands. Defence and alarm substances are thus in this case always, of necessity, disposed of together. This is, however, not the case with other groups of ants. Cavill and Williams (1967), as well as Hermann and Blum (1967) have shown that for Myrmeciines, Pseudomyrmecines and Myrmicines the contents of both glands may also be disposed of separately.

A close connection between alarm stimulation and defence can also be demonstrated in the case of the metathoracal gland, which is present in all ants, apart from a few parasitic species. The secretion of this gland is discharged through a reservoir. It was originally believed that the secretion of this gland produced a substance peculiar to each species, which served for purposes of recognition of members of the same species and thus also for the differentiation of ants of other species (Janet, 1898; Brown, 1968). It was possible to demonstrate, however,

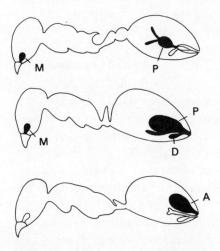

Fig. 5.1   Alarm substance gland (shown in black) in different species of ant. Above – *Myrmica;* centre – *Formica;* below – *Tapinoma* (Dolichoderine). A. anal gland; D. Dufour's gland; M. mandible gland; P. poison gland (after Hölldobler, 1970b, adapted).

that in all hitherto researched species of the subfamilies of the Myrmeciines, Ponerines, Myrmicines, Pseudomyrmecines and Dolichoderines the secretion inhibits the growth of microorganisms (Maschwitz *et al.,* 1970; Maschwitz, 1974). Among certain known species the metathoracal gland is greatly enlarged and serves to ward off enemy animals. *Crematogaster difformis* belongs to this group of species; in this species the glandular secretion is stored in a greatly widened reservoir and smells strongly of phenol. In other ant species this secretion has a strong deterrent effect (Maschwitz, 1974). In *Crematogaster inflata,* one of the species, like *Crematogaster difformis,* occurring in Malaysia, the metathoracal glandular secretion does not

have a repellent effect on other ant species. It is nevertheless used as a
defence substance. It can in certain cases be actively secreted and stuck
to the extremities of the opponent, owing to its sticky consistency (*see*
pp. 260–61). Apart from this it also serves as a means of communication.
It has the effect of drawing and attracting workers of the same species.
In the experiments conducted by Maschwitz, the attracted workers bit
and tore at the piece of filter paper which had been smeared with the
glandular secretion. This shows that this secretion also triggers aggressive
behaviour and serves as an alarm substance.

A great deal is known about the chemical composition of the alarm
substances of some species, but as yet, nothing is known of those of
other groups, such as the Dorylines. We know only from the
experiments of Brown (1960) that a strong alarm response can be
triggered by the extract from the mandible gland in certain species of
army ant, such as *Eciton, Labidus* and *Nomamyrmex*. Robertson (1971)
described how, for the Myrmeciines, aggressive behaviour can be
triggered by the secretion of the mandible gland, Dufour's gland and the
contents of the lower intestine in the case of *Myrmecia gulosa*. Apart
from this, we know from the work of Cavill and Williams (1967) that
the main components of the Dufour gland secretion are
cis-8-heptadecene and n-pentadecane. From the Ponerine group,
Paltothyreus tarsatus was investigated in more detail by Casnati *et al.*
(1967). They found that in this case the main components of the alarm
triggering substance of the mandible gland were dimethyldisulphide and
trimethyldisulphide.

Of the remaining three subfamilies, a great many more alarm
substances have been chemically analysed and to a great extent
compiled by Blum (1969). The characteristic feature of these substances
is that they contain structurally relatively simple terpenes, ketones,
aldehydes, alcohols and alkanes, with a molecular weight of between
100 and 200. This characteristic molecular weight, which is low for
alarm substances, brings them well in line with their function under
natural circumstances.

Alarm substances chiefly serve the purpose of indicating disturbances
in the area of the nest. It is thus important that the communication
should occur as speedily as possible, in order to alert as many members
of the nest as is necessary to eliminate the danger, but that, on the other
hand, will not alert too many workers and mobilize the entire colony
for every minor disturbance. Lastly, the alarm affect should fade as

quickly as possible after the danger is ended. It is thus necessary for an optimal chemical alarm to be spatially and temporarily as precise as possible. In fact Wilson and Bossert (1963) have discovered that in the harvester ant species *Pogonomyrmex badius* distributed in the USA the scent of the alarm substance of a single worker extends to a maximum of 6 cm. The nest mates are alerted inside 13 s within this radius; the alarm effect fades after 35 s. If the disturbance in the nest increases, however, then the number of alerted ants also increases and so does the amount of alarm substance released; the distance across which the scent can travel and the length of duration of the alarm increase proportionately.

The speedy action of the alarm substances and their spatial and temporal localization are achieved due to the fact that the molecules of the alarm substance are relatively small, as a result of which the pheromones evaporate quickly. On the other hand, the ants' smell receptors are not very sensitive. The concentration threshold of these receptors in ants, in comparison with the receptors for sexual scent in butterflies, is quite high. This emerges from behavioural experiments conducted by Moser *et al.* (1968) with the leaf-cutter ant *Atta texana* (alarm threshold 3×10^8 molecules per cm^3).

Alarm substance receptors have also been successfully investigated with electrophysiological methods (Dumpert, 1972a). These receptors are situated in the sensilla trichodea curvata on the feelers. Of the large number of sensory cells in these sensillae (over 30), a specific type react to the alarm substance of this species, undecane. The minimum concentration of undecane which one must present in order to obtain a reaction from this cell type is 5×10^{10} molecules per cm^3. If one presents other substances with molecules containing carbon chains of only 10 (decane) or 12 (dodecane) components, then the concentration threshold is raised by three powers, in other words a thousandfold. Substances differing from undecane by two links (nonane and tridecane) have no effect whatsoever. This shows that the sensory cells which react to the alarm substance are able to distinguish very precisely between undecane and homologous substances related to it.

Up to the present time, the alarm pheromones of about 1% of all known ant species have been proven to exist and analysed. This has shown that alarm pheromones are usually not species specific, but specific to the larger systematic groups of species. Undecane was thus found in the Dufour gland of all hitherto researched species of Formicine, whilst for the Myrmicines specific ketones are counted as typical alarm substances. In two cases known today, the alarm substance of the ant species is chemically identical with that found in bees. One example of this is 2-heptanone, which is produced in the mandible gland of the honey bee (*see above*), and which has also been demonstrated in a few species of the subfamily of the Dolichoderines (*Azteca spec., Dorymyrmex pyramicus* and *Iridomyrmex pruinosus*) as being an alarm substance (Kannowski *et al.*, from Blum, 1969; Blum and Warter, 1966). It is frequently the case that not just one but several substances may be produced as alarm pheromones by one species. One is able to see this from the fact that species from such groups as the Myrmicines and Formicines have several glands in which they produce alarm substances. Several substances may even be produced for alarm purposes within the same gland. If one then compares the proportions of these substances produced in the same gland in various related species, then the differences shown in *Myrmica* are such that one can clearly distinguish the different species from these proportions (Crewe and Blum, 1970). Thus, although the various components may trigger an alarm separately, it may nevertheless also be the case that these various substances must be presented in proportions specific to the species in order to obtain the optimal stimulus for alarm behaviour.

Glandular secretions may also vary in composition within the same species, and even among different individuals of the same colony. Tricot *et al.* (1972) found in the case of *Myrmica rubra* that the main components of the mandible gland secretions, 3-octanol and 3-octanon, could be very differently composed in different workers of a colony. The proportion of 3-octanol lay in the region of 4.6 and 30%. According to the observations of Tricot and his colleagues, in *Myrmica rubra* it is the poison gland secretion that triggers alarm behaviour, whilst 3-octanon on the other hand attracts other workers, at the same time suppressing aggression. Tricot concludes from these observations that both pheromones work together in the case of an attack against these ants. The secretion of the poison gland triggers agressive behaviour in members of the colony in the immediate vicinity of the place of danger, whilst the 3-octanon from the mandible gland attracts further workers, at the same time as its aggression-inhibiting effect precludes the possibility of the ants giving the alarm being attacked by those receiving it. 3-octanol, on the other hand, shows no direct effect on the behaviour of the ants, but 3-octanol and 3-octanon evaporate at different speeds; this may assist the *Myrmica* workers in reaching the place of danger

more easily. However, this supposition by Cammaerts-Tricot (1973) has
not yet been demonstrated.

In the case of the African weaver ant *Oecophylla longinoda,* it has
been found that several substances work together in raising the alarm. In
the alarm-triggering secretion of the mandible gland Bradshaw *et al.*
(1975) found over 30 different substances, including hexanol, l-hexanol,
3-undecanone and 2-butyl-2-octenal. When these four substances were
individually tested for their effects on the behaviour of the weaver ants,
it was shown that hexanol triggered an undirected alarm reaction in the
ants, whilst l-hexanol on the other hand led the alerted animals in the
direction of the source of the scent and 3-undecanone and 2-butyl-2-
octenal triggered a biting reaction. These various reactions were not
simultaneous with the natural secretion of the mandible gland, but
staggered and triggered at varying distances from the source of the scent.
This is very probably due to the fact that the various secretions of the
mandible gland evaporate at different rates, or that the ants react to
these substances at varying thresholds. The various substances of the
mandible gland, as emerges from the experiments of Bradshaw and his
colleagues, are not produced by all *Oecophylla* workers in either equal
quantities or composition, and in any case do not all react in the same
way to the various components. For example, l-hexanol affects the
small workers as a repellent and disperses them, whilst it attracts the
larger workers to follow them. Apart from the secretion of the mandible
gland, formic acid from the poison gland and undecanone from the
Dufour gland also act as alarm substances; finally, it would appear that
in the case of *Oecophylla,* the tapping signal against the walls of the nest
is also for the purposes of raising the alarm. In what way these alarm
signals and the as yet, unresearched components of the mandible gland
substance work together, remains a mystery.

In the case of the wood ant *Formica rufa,* the formic acid already
mentioned is not the sole alarm substance either; apart from this, the
secretion of the Dufour gland also triggers alarm behaviour, and is given
off automatically with the substance from the poison gland. Analysis of
the secretion of this Dufour gland has revealed that it contains 22
substances, nearly all of which belong to the simple saturated or
unsaturated hydrocarbonates. If one compares the proportions of these
various components, undecane occupies the largest place by over 50%,
followed by tridecane at 20% (Bergström and Löfqvist, 1973). The
behavioural experiments of Löfqvist (1976) show, as do the
electrophysiological experiments of Dumpert (1972a), that the formic

acid of the poison gland and the hydrocarbonates of the Dufour gland stimulate different types of receptor cells. The simultaneous stimulation of the various receptor cells by a mixture of formic acid and the hydro-carbonates of the Dufour gland triggers a more intense and longer-lasting alarm response than formic acid alone (Löfqvist, 1976).

In some ant species, the concentration of the alarm pheromones also plays an important part. Varied concentrations of the alarm substance of the same species at times alter not only the intensity, but also the nature of the reaction. Wilson (1958b) found with the harvester ant *Pogonomyrmex badius* that the secretion of the mandible gland had the effect of attracting the ants in small quantities, but triggered aggressive behaviour in greater concentrations. If the ants are presented with the substance for any length of time, they begin to dig. This is similarly the case with some Dolichoderine species, whose alarm pheromones bring about an attraction response in small quantities, but in greater concentrations on the other hand similarly stimulate aggressive behaviour (Wilson, 1963). In the case of *Tapinoma sessile,* which also belongs to the Dolichoderines, high concentrations of the substance, if presented for a sufficient length of time, lead to the complete evacuation of the colony from the nest (Wilson, 1963).

A particular group of alarm pheromones were designated 'propaganda substances' by Regnier and Wilson (1971). These substances were found in very large quantities (1/10 of the total body weight) in the form of acetates (decyl-, dodecyl- and tetradecylacetates) in the particularly large Dufour glands of the slave-making ants, *Formica pergandei* and *Formica subintegra.* We shall be discussing the slave-making ants in more detail later. Anticipating this sufficiently for present understanding, we can say here that slave-making ants invade the nests of other specific species and steal pupae. The stolen pupae then mature in the nest of the slave-makers, and take over specific tasks in this nest. The propaganda substances are placed by the groups of robbers seeing to the convoy of new pupae. They are given off in the conquered nest and have the effect of attracting the robbers and frightening off the workers of the alien species, who flee in panic. The effect of these substances is evidently long-lasting, since the ants who have been attacked only dare to return to their nest after some days. Inside the slave-makers' own nest, these propaganda substances act as alarm pheromones. According to the accepted definitions these acetates of the Dufour gland are allomones as well as pheromones.

The beetle species *Pella japonicus* (Sharp) and *Pella comes* (Sharp), belonging to the Staphylinids, show how it is possible to beat ants at their own game. They intermingle with workers of *Lasius spathepus*, (native to Japan) out on a food search and eat on the way not only dead ants, but also small insects being carried back by the ants. These beetles are generally tolerated by the ants. If they are once attacked, however, they release a substance which has a very similar effect on *Lasius spathepus* as their own alarm substance (citronella). The *spathepus* workers immediately stop and change their course. The beetles make use of this reaction by making a quick exit (Kistner and Blum, 1971).

5.3   Recruitment

Recruitment plays an important part in the changeover to a new nest, as well as in the exploitation of new sources of food. In both cases, the members of the nest have to be brought to leave the nest (recruitment) and must, moreover, be informed as to the position of the new nest or feeding place as appropriate. The precondition for successful recruitment in both cases is communication between the scouts and their nest mates.

   One of the original, if however uneconomical, methods of recruitment is presented by the 'tandem travel' method, whereby the nest mates are led individually to the feeding place or the new nest. Hingston (1929) has observed this form of behaviour with *Camponotus sericeus*, Wilson (1959) with *Cardiocondyla venusta* and *Cardiocondyla emeryi* and Dobrzanski (1966) with *Leptothorax acervorum*, a common European species. The feature held in common by all these species is that their colonies are relatively small. Hölldobler, Maschwitz and Möglich have researched into the 'tandem travel' of *Bothroponera tesserinoda, Camponotus sericeus* and *Leptothorax acervorum* more precisely and have been able to clarify the signals by which this behaviour is stimulated.

   In the case of *Bothroponera tesserinoda*, a tropical ponerine with between 50 and 200 animals per nest, the scouts returning from the new site of the nest or a feeding place display a characteristic summoning behaviour: they go to individual nest mates and take hold of their heads with their mandibles. They then immediately turn round, thus presenting the rear end of their bodies to the nest mate being summoned to come with them. If the summoning has been successful, the summoned ant then joins on to the ant leading, inducing her to continue

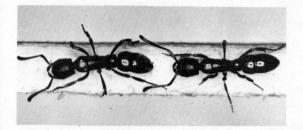

Fig. 5.2   Tandem travel of *Bothroponera tesserinoda* (Photo – U. Maschwitz).

travelling. The summoned ant follows the scout, feeling the hind part of its body and its hind legs with its antennae four to six times per second. She occasionally also nudges the rear end of the leader with her head. The pair remain together in this manner, usually until the feeding place or the nest site has been reached. However, if the animals lose each other in the meantime, then the leader first stands and waits for about four seconds; she then moves round in circles, as does the ant being led before her, and searches for the lost partner. In this way, the tandem pair usually find each other quickly and continue on their way (Hölldobler *et al.,* 1973; Möglich, 1973; Maschwitz *et al.,* 1974).

The signal which brings about the leading behaviour of the recruiting ant is that of the mechanical, repeated contact of the ant being led, through its head and feelers. This emerges from subsequent research by Maschwitz *et al.* (1974): a tandem pair were separated by slipping a small piece of paper between them. The leader thereupon stood still and waited, whilst the ant being led showed the search behaviour pattern. If the leader was then touched at least twice with a hair on its hind legs and the rear end of its body, then it continued on its former course.

The ant being led, on the other hand, is caused to follow the leading ant by both a mechanical and a chemical signal, either of the two signals being effective on their own, but weaker than a combination of the two. The mechanical signal is the rear section of the leading ant's body, which can, for example, be substituted by means of a glass or wax pellet from 1 to 9 mm in diameter and in the appropriate colour. The chemical signal is a surface pheromone, which cannot be linked to any known pheromone gland. It is probably not released only during tandem travel, but is found at all times on the body surface of all animals in like manner on the head, thorax and rear section. It may possibly be identical with the 'colony scent' of the ants (*see below*), since the 'guiding properties' of artificial traps increase in the case of *Bothroponera* by being kept inside the nest for a few weeks rather than days (Hölldobler *et al.*, 1973).

The tandem travel of *Camponotus sericeus* (Formicine) differs from that of *Bothroponera tesserinoda,* first in that all parts of the body do not act as equal stimuli to the ant's reaction. The rear section of the body is more effective in this respect than the head or thorax. Apart from this, the chemical signals of the leading ant are also necessary for a tandem to come about. The *Camponotus* scouts also lay a scent trail from the new nest site or feeding place to the nest. This trail in itself has no recruiting effect, but merely serves the scout as an aid to orientation during the tandem course. The trail is laid with the content of the rectal bladder, the widened end of the intestine. Finally, in contrast with *Bothroponera,* the summoning behaviour of *Camponotus sericeus* is different for food recruitment than for a change of nest. Whilst the *Camponotus* scouts when summoning others for a change of nest, like the *Bothroponera,* draw the nest mates to themselves and then turn through 180°, when enlisting for their feeding place they first go round the nest with a full crop, displaying exchange of food and cleaning behaviour patterns. After the scouts have repeatedly offered food, they finally offer it only 'ritually', inducing individual nest mates to follow after them, and thus bringing about the formation of a tandem (Möglich, 1973).

Finally, *Leptothorax acervorum,* the third species researched more precisely for its tandem travel (Möglich *et al.,* 1974), displays a type of behaviour which deviates from the others, particularly in the method by which the ants are summoned. A scout, having discovered a new food source, offers its food and gives it to the nest mates, like the *Camponotus* scout. She then turns round, however, and stands with the rear section of her body raised, and extending her sting on which a small droplet of liquid can be seen ('tandem ventilation'; *see* Fig. 5.3). The droplet of liquid comes from the poison gland, whose secretion exercises an attraction effect on the nest mates. As soon as the first nest mate reaches the scout and touches her with her antennae on her hind legs and the rear section of her body, the scout lowers the hind part of her body and leaves the nest. The tandem has been formed and will remain together by means of the feeler touches of the ant following and the presented secretion of the poison gland of the leader.

According to the research of Möglich *et al.* (1974), other *Leptothorax* species, such as *Leptothorax muscorum* and *Leptothorax nylanderi,* display the same behaviour during tandem travel, with only the one difference, that the rear section of the *Leptothorax nylanderi* body is not raised quite so high during tandem ventilation as with the other species. The active pheromone of the poison gland secretion in tandem ventilation is only partially species-specific. *Leptothorax acervorum* thus reacts to the pheromone of *Leptothorax muscorum,*

Fig. 5.3   Tandem ventilation (a) and tandem travel (b) of *Harpagoxenus sublaevis* (Photo — A. Buschinger).

but not to that of *Leptothorax nylanderi.* The latter species reacts only to its own pheromone.

The slave-making ant *Harpagoxenus sublaevis* recruits its *Harpagoxenus* nest mates by tandem travel for predatory expeditions, during which pupae are stolen for future slaves from the nests of *Leptothorax acervorum* and *Leptothorax muscorum.* The summoning to tandem travel ensues when the returned scout vigorously touches her nest mates with her feelers and front legs, also performing 'shaking motions'. After these signals on the part of the scouts, the formation of a tandem usually follows, or alternatively a tandem ventilation at the entrance to the nest (Fig. 5.3), which similarly serves to enlist for the tandem. When the tandem has reached its objective, the pair then separates by means of a specific 'disconnecting behaviour', in which the

leader of the tandem either jumps free in one leap, or turns toward the
animal being led and touches it with her feelers (Buschinger and Winter,
1978). According to the observations of Buschinger and Winter, there
are also mixed tandems, in which *Harpagoxenus* take their slaves,
ill-suited for fighting, on predatory expeditions, and also tandems in
which *Leptothorax* workers lead their slavemakers to feeding places.

The characteristic feature of the tandem travelling of *Leptothorax*,
as against those of *Bothroponera tesserinoda* and *Camponotus sericeus*,
consists in the fact that *Leptothorax* does not deposit a non-specific
surface pheromone, but a specific pheromone such as is employed by
other species for laying trails, or as sexual pheromones. The slave-maker
ant *Harpagoxenus americanus* widespread in the USA actually leads its
nest mates recruited to the tandem to the overtaken ants by means of a
scent trail (Wesson, 1939), whilst the connection between the
pheromone deposited during tandem ventilation and the sexual
attraction substances consists in the fact that in *Leptothorax*,
Harpagoxenus and a few other species females ready to mate attract the
males by means of an 'attraction ventilation' (*see below*), inducing them
to copulate. The secretion of the poison gland, which contains a sexual
attraction substance, is also presented during attraction ventilation.
Amongst other arguments, this supports the hypothesis of Hölldobler
(1973d), according to which among at least some of the Myrmicines,
sexual attraction substances and recruitment pheromones had an
identical phylogenic source.

The nest mates of *Formica fusca* are recruited for removal into
another nest in a very simple manner, comparable in some respects with
the tandem method. They are simply carried there. The summoning
behaviour for carrying is very similar to that for the tandem. Here also
the scout seizes a nest mate by the mandibles, draws it towards her and
then turns through 180°. In this case, however, she does not release the
nest mate, but turns the ant around with her. The usual reaction of an
ant seized in this way is for her to tuck the rear section of her body
underneath herself and draw her extremities in close to her, so that she
can be carried (Fig. 5.4).

Among some *Leptothorax* species, both carrying behaviour and
tandem travelling occur together, during a nest changeover. Some
workers of the *Leptothorax acervorum* species, probably inexperienced
animals working inside the nest, display a summoning behaviour 'to be
carried' related to the carrying behaviour pattern. They lie inclined to
one side and are given priority in being carried by their nest mates

(Zebitz, 1979). One finds similar carrying behaviour in connection with removal into another nest in other species of *Formica*, such as *Formica polyctena* and *Formica sanguinea*. Among the carpenter ants *Camponotus herculeanus, Camponotus ligniperda* and among other *Formica* species, chemical signals also seem to play a part in the removal to a new nest, ancillary to the carrying behaviour (Möglich, 1971).

In the cases of *Formica fusca, Formica sanguinea* and *Camponotus sericeus* there are, according to the research of Möglich, only a few workers evidently, who organize the complete removal into the new nest, whilst the rest of the colony members are passively transported or led, as the case may be. It emerges from subsequent research that if one removes the carrying or leading animals from tandem pairs, the removal into the new nest quickly comes to a halt. If one presents a species such as *Formica fusca* with two different new nests, allowing the two nest sites to be discovered simultaneously by different scouts, the colony

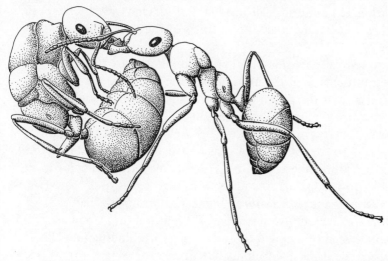

Fig. 5.4   Formica rufa worker carrying a queen (after Möglich and Hölldobler, 1974, adapted).

then becomes split: the two groups of scouts transport the remaining nest mates into 'their' nest.

The recruitment of nest mates by means of carrying among *Formica polyctena* does not only play an important role in removal to a new nest. It is also important, as Kneitz (1964) has observed, in the exchange of nest mates between different nests, the changeover from the summer nest into the winter nest and vice versa and also in the

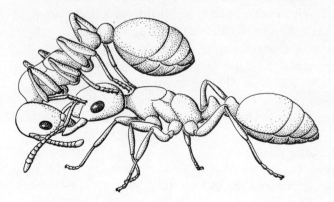

Fig. 5.5   Leptothorax nylanderi worker carries another worker (after Möglich and Hölldobler, 1974, adapted).

transportation of young animals inside the nest. One can observe peak periods of carrying activity at certain times of the year, i.e. autumn and spring.

Nest mates are also carried in other subfamilies apart from the Formicines; in the case of the Myrmicines, the ant being carried is taken by the mandibles or the throat and then holds her thorax and gaster behind her, above the body of the carrying ant (Fig. 5.5). Ponerines carry their nest mates in the same way as they carry other things, namely parallel to the longitudinal axis between their legs (Maidl, 1934; Maschwitz and Mühlenberg, 1973b). The role played by carrying in the case of fully grown nest mates still requires further research.

In contrast perhaps with *Bothroponera, Formica fusca,* like *Camponotus socius,* recruits nest mates for removal into a new nest and for the exploitation of a new food source in a different manner. The *Formica fusca* scout, having found a new food source, lays a trail between the feeding place and the nest with the secretion of the rectal bladder. This trail serves only as an aid to orientation and has no stimulative effect on the nest mates. Only a specific rocking behaviour of the scout which laid the trail, and the passing on of the food brought with her in her crop induces other workers to leave the nest and follow the trail laid by the scout. An orientation trail is also laid with the contents of the intestine in the case of the recruitment of nest mates for removal into a new nest; the trail becomes more marked by scouts travelling repeatedly from the old to the new nest. The scouts then enter the old nest and summon the nest mates with a specific behaviour pattern to leave the nest and follow the scent trail to the new nest. This summoning behaviour consists of the scouts displaying short, swift 'sprints' in the nest, then running to the nest mates and rocking evenly forwards and backwards after an initial shaking motion. Food exchange – in contrast with feeding recruitment – hardly ever occurs. Instead of doing this, the scouts usually grip the animated nest mates with their mouth parts, draw them towards themselves and carry them to the new nest. The majority of the colony members are transported in this fashion, whilst only a few workers are induced to leave the nest independently by means of the shaking and rocking behaviour Möglich and Hölldobler, 1975).

In the case of *Camponotus socius*, the scouts also lay a trail with the contents of their rectal bladder from the feeding place to the nest. This trail secretion too has only an orientation function. For the recruitment of the nest mates, it is necessary for the scouts to display shaking behaviour, during which they move the head and thorax from side to side 6-12 times per second (for about 0.5-1.5 s); this signal alone induces up to 30 workers to follow one scout and leave the nest. The scout then leads the summoned ants to the feeding place, following her own trail as an orientation aid. If the leader is taken away, the summoned ants do not travel further. This shows that further signals from the leader are required on the path to the feeding place, signals which were in fact traced to the poison gland. Hölldobler was able to prove this by ejecting a mixture of the contents of the rectal bladder and the substance of the poison gland from a microsyringe, which induced the 'leaderless' *Camponotus* workers to continue travelling. In this experiment, as in natural recruitment, the secretion of the rectal bladder serves as a long-term orientation signal, whereas the contents of the poison gland act as a short-term summoning signal, stimulating the ants to follow the trail (Hölldobler, 1971b).

In the case of *Camponotus socius,* the recruitment of nest mates during removal into a new nest depends on essentially the same signals as feeding recruitment. One is only able to recognize differences in the behaviour of scouts in the nest. During removal into a new nest, the colony members are not summoned by the shaking behaviour, but by the scouts moving their heads forwards and backwards. By means of this signal, not only workers but also males are summoned to leave the nest. Only such nest mates as do not react to the summoning behaviour of the scouts are carried to the new nest.

Within the Formicine group, among those species which have
subsequently been investigated, the trail secretion alone effects the
recruitment of the nest mates. Subsequent investigations showed that it
required no further signals on the part of the scouts to stimulate the
nest mates and induce them to follow the trail. It is a universal feature
of all Formicines that the trail pheromone is stored in the rectal bladder
and released through the anus (Blum and Wilson, 1964; Fig. 5.6). The

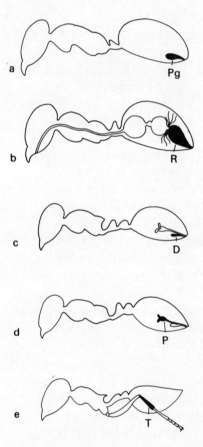

Fig. 5.6   The trail substance glands of various ant genera: *Tapinoma* (a);
Lasius (b); *Tetramorium* (c); *Solenopsis* (d); *Crematogaster* (e). D. Dufour's gland;
P. poison gland; PG. Pavan's gland; R. rectal gland; T. tibial gland (after Hölldobler,
1970b, adapted).

chemical composition of the trail substance of Formicines is, up to now, only known in the case of *Lasius fuliginosus*. The scent trails of this species are particularly easy to follow; one is able to see the course they take in a nearly unbroken chain of workers moving either to or from the food source. It is, therefore, hardly surprising that the trail laying of this species has been particularly exhaustively researched.

In 1914, Brun made *Lasius fuliginosus* workers walk over a piece of sooted paper, and was able to detect under a magnifying glass not only the ants' trail of footprints, but also clearer lines, which he took to be the result of active marking by the ants. Brun thus reinforced the observations of Santschi (1911), who saw how various ant species dotted the ground with the tip at the rear section of their bodies. Stumper (1921a) then succeeded in laying trails with the liquid extract of whole *Lasius fuliginosus* animals, the trails proving effective in behavioural tests; Carthy (1951b) was able to give the first indications of the source of the trail substance. He fed charcoal and staining solution to workers, and having established that both substances reappeared in the trail, concluded that the trail substance was released through the anus. This was later proved by Hangartner and Bernstein (1964), by laying trails with the diluted contents of the rectal bladder; the trails proved to be behaviourally effective.

By using this method of laying artificial trails with the diluted contents of the rectal bladder, Hangartner (1967 and 1969a) was able to clarify certain questions as to the specifics of the trail substance, the stability of the scent trails and the orientation of the workers.

It was first demonstrated that the trail substance of *Lasius fuliginosus* is not effective among other *Lasius* species, such as *Lasius emarginatus, Lasius niger* and *Lasius flavus,* but that on the other hand *Lasius fuliginosus* is well able to follow the trails of *Lasius emerginatus* and *Lasius niger. Lasius flavus* seems, by contrast, to produce no trail substance at all, or if so one with only minimal effectiveness.

Since the work of Huwyler *et al.* (1973 and 1975), a point has been reached in the chemical analysis of the trail substance of *Lasius fuliginosus* where we are already able to recognize six of its component substances. We are dealing in all cases here with fatty acids, i.e. hexane, heptane, octane, nonane, decane and dodecane acids. It is interesting to note here that one of these acids, hexane, has also been proved to be present in the trail substance of a termite species, *Zootermopsis nevadensis* (Hummel and Karlson, 1968).

Hangartner's observation, that the underground species *Lasius flavus* produces no trail substance, or only a minimally effective one, made it reasonable to suppose that underground ant species lay no trails in their underground passages. Hangartner, in order to discover whether this was so, investigated the underground species *Acanthomyops interjectus,* which is superficially similar to *Lasius fuliginosus,* but found that this species also lays trails.

Trail laying is similarly widely found in the Dolichoderines and Myrmicines. As with the Formicines, the odour trail serves in the majority of instances not only for the orientation of the ants, but also at the same time for the recruitment of nest mates. In certain cases, however, the two functions are separated. As with *Camponotus socius,* these involve essentially mechanical signals, by means of which the scouts stimulate their nest mates, whilst the odour trails mainly serve purposes of orientation. Examples of this are known in species of *Monomorium* and *Tapinoma* (Szlep and Jacobi, 1967) and also *Pheidole* (Szlep-Fessel, 1970).

Among *Monomorium (Monomorium subopacum* and *Monomorium venustum),* the scout lays a poison gland trail from the food source to the nest, using its sting; this does not, however, activate any other workers by itself. The recruitment of nest mates occurs only through the behaviour pattern of the scouts running to their nest mates inside the nest, nudging them with their heads and feeling them with their antennae. The recruited workers leave the nest and follow the trail laid from the feeding place to the nest by the scouts, who are no longer required.

Recently a number of the components of the *Monomorium pharaonis* trail substance have been successfully analysed. These are 5-butyl, 3-methyl-octahydroindolizine; 2-butyl, 5-pentyl-pyrrolidine and 2-(5-hexenyl), 5-pentyl-pyrrolidine, which have been designated as Monomorin I, II and III, respectively (Ritter *et al.,* 1973, Ritter *et al.,* 1975). Apart from these, a bicyclical unsaturated hydrocarbonate of the summation formula $C_{18} H_{30}$ was found, and was named Monomoren. The Monomorines can nevertheless also be artificially produced (Oliver and Sonnet, 1974; Ritter and Stein, 1975). For a long time, however, we have been unable to discover anything about the biological function of the various chemicals which comprise the trail substance.

Among *Pheidole pallidula* and *Pheidole teneriffana,* the same features are to be found during the recruitment of nest mates as with *Monomorium subopacum,* i.e. running, nudging and touching with the antennae. In the case of *Pheidole,* however, a proportion of the workers are induced to leave the nest as a result of the trail laid by the scouts, without requiring any further stimulus, whilst the soldiers present among this species are recruited mainly by the summoning behaviour of the scouts (Szlep-Fessel, 1970).

Of particular interest is the feeding recruitment of *Crematogaster* species, which also belong to the Myrmicines. Species of this genus, recognizable by the characteristic shape of the rear section of their bodies which is raised upwards in an obtuse angle when the ants are aroused (Fig. 5.7), are found only in warmer climates, such as the

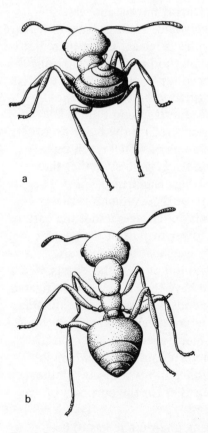

Fig. 5.7   Crematogaster peringueyi worker laying (a) and following (b) a trail (after Fletcher and Brand, 1968, adapted).

Mediterranean. *Crematogaster* workers also display a particular behaviour pattern with which they recruit nest mates (Leuthold, 1968a). The orientation of the recruited workers follows by means of an odour trail, laid by the scouts between the feeding place and the nest. The trail secretion, however, is not produced in the rectal bladder as in the case of the Formicines, or in the poison gland as with the hitherto named Myrmicine species, but is produced in the hind legs (Fig. 5.6). The tendons responsible for the movement of the claws are so far apart in the tibia of the hind leg that they form a cavity in which the trail substance is stored. The pheromone travels inside the tendon from this point, and comes out of the first foot section. The trail substance can be actively released by the ants, since the hind legs are not coordinated with the two front pairs of legs, but are able to trammel the ground rapidly, one after the other (Fig. 5.7). During normal movement, however, when a trail is not being actively laid, all three pairs of legs are coordinated together. Some trail substance is released even then, but far less than during active trail laying. This method of trail laying using the hind legs, which is unusual for ants, has so far been discovered to exist in two *Crematogaster* species, *Crematogaster ashmeadi* (Leuthold, 1968a,b) and *Crematogaster peringueyi* (Fletcher and Brand, 1968).

In *Melissotarsus titubans* recently discovered on the Ivory Coast, there are widened tarsal glands, the products of which reach the outside of the body via a number of vessels. It is possible that these glands also produce a trail substance (Delage-Darchen, 1972a). The remarkable, and for ants unique, feature of these animals is their method of locomotion, for they live only in passages under the bark of trees. They walk with four legs on the lower surface of the passage, but with the remaining two legs on its upper surface: these ants are incapable of moving in a coordinated fashion outside their nests.

Amongs the other Myrmicines which have thus far been investigated, the trail pheromones are produced in glands belonging to the stinging apparatus, that is the poison gland, or alternatively the ancillary or Dufour's gland which we have already mentioned (Fig. 5.6). In the cases of *Pogonomyrmex badius,* the harvester ant species which is widespread in North America, and *Myrmica rubra,* a species abundant in central Europe, both glands work together for the purposes of feeding recruitment.

In *Myrmica rubra,* the actual recruiting pheromone has to be produced in Dufour's gland, whilst the poison gland supplies a pheromone serving for the orientation of the ants only. A *Myrmica rubra* worker, having discovered a spoil and unable to deal with it single-handedly, marks the given area around it with the secretion of the Dufour's gland. The worker then goes back to the nest, laying a trail with the contents of the poison gland. Coming away from the nest again, as has been stated, the worker lays a trail with the secretion of the Dufour's gland, thus attracting the nest mates to follow (Cammaerts-Tricot, 1974a,b).

In the case of the harvester ant species *Pogonomyrmex badius* and also a range of other *Pogonomyrmex* species, the functions of the glandular secretions are exactly the other way round. Here the poison gland contains a recruitment pheromone and the Dufour's gland a pheromone for the purposes of the orientation of the workers (Hölldobler and Wilson, 1970). Chemical analysis of the contents of the Dufour's gland of *Pogonomyrmex badius* showed that the secretion comprises various branching and non-branching hydrocarbons, with dodecane and 6-methyldodecane as the main components. It may be that the chemical composition and combination of these substances is not only specific to the species but also even to the colony. This idea is supported by initial chemical analyses and behavioural experiments, which indicate that *Pogonomyrmex badius* workers are able to distinguish their own trails from those laid by *barbatus* workers of other colonies (Regnier *et al.,* 1973).

The secretion of the Dufour's gland of certain *Pogonomyrmex* species, such as for example *Pogonomyrmex barbatus* and *Pogonomyrmex rugosus,* can be used to lay lasting trails, which remain in existence for long periods and are even able to resist heavy rainfall. These lasting trails, whose 'odour lines' are probably variable according to specific colonies, structure an area inhabited by a number of species. These lasting trails run in such a manner that the trails of different colonies do not cross over each other. Since the majority of workers of a particular colony travel along the lasting trails in search of food, and similarly travel back to the nest along them, the ants are thus channelled in such a way that relatively few meetings, and thus also fights between members of different colonies arise. This dividing-up of feeding places also facilitates a greater density of nests in a given area: one can deduce this from the fact that the nests of *Pogonomyrmex maricopa*, a species which lays no lasting trails, lie more than twice as far apart as those of *Pogonomyrmex rugosus* and *Pogonomyrmex barbatus* (Hölldobler, 1974, 1976a).

Among *Pogonomyrmex rugosus, Pogonomyrmex californicus* and

other harvester ants, other means ensure that workers of different species encounter each other very little whilst searching for food; they search for food at different times of day, different ants being activated in a species-specific way to various temperatures of the surface of the ground (Clark and Comanor, 1973; Bernstein, 1974).

The fire ants (*Solenopsis saevissima* group), widely found in the southern states of the USA and feared on account of their painful sting, lay trails with the extended sting, using the contents of the Dufour's gland, as emerges from the investigations of Wilson (1962b,c) (Fig. 5.6). This trail substance serves both for the recruitment and the orientation of the nest mates. We deduce this from the fact that the proportion of recruited workers increases in direct ratio to the amount one increases the quantity of trail substance on an artificially laid trail (Wilson, 1962b). It was therefore interesting to discover whether individual workers were able to match the number of workers going to feed with the quantity or quality of the food found, by releasing more or less trail substance as appropriate. Wilson (1962c) however was unable to find any such correlation between the trail laying of individual ants and the quantity and quality of the food source. Trail laying occurs on an 'all-or-nothing' basis. They are either laid or not laid; there do not seem to be any graduated differences. We had to assume from these results that the number of ants recruited may only be matched to a particularly rich food source in such cases where the trails are made more intense by the number of returning ants renewing them.

Hangartner (1969c) later renewed the investigation into this question, and found that the trails of individual workers in fact can be variably deposited (Fig. 5.8). The continuity of the trail and at the same time, the pressure of the sting, increased in the fire ant investigated by Hangartner, according to increased need for food on the part of the colony, increasing quality of the food source and decreasing distance between the feeding place and the nest. It was not possible to ascertain with the fire ant, however, whether variably deposited trails also corresponded to variable amounts of trail substance being released. The amount of trail substance released by a fire ant is too small to allow this to be possible. To show how small the quantity actually is, Walsh *et al.* (1965), estimated the amount of trail substance produced by a *Solenopsis saevissima* worker as being 0.6 ng, that is 0.6×10^{-9} g. A more suitable animal for experiment in this respect was the Formicine, *Acanthomyops interjectus,* which deposits individual odour marks when laying a trail. Hangartner was able to show here that the quantity of

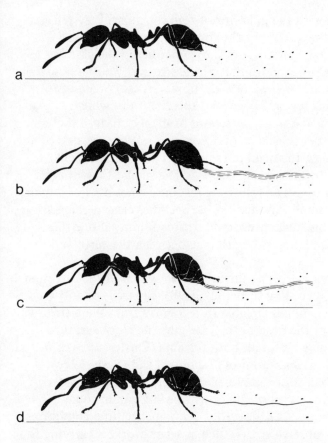

Fig. 5.8   Solenopsis geminata worker, which lays variable trails on a sooted glass plate, depending on the abundance of the food source. (a) foot trail only for a small food source; (b) trail with the tip of the rear section of the body; (c) trail made with the bristles and lightly with the sting; (d) strong trail made with the sting for a very abundant food source (after Hangartner, 1969c and Hölldobler, 1970b, adapted).

trail substance released by individual workers also corresponds to the quality of the food source. The individual ant deposited denser odour marks, depositing more trail substance per mark, with increasing concentrations of the sugar solution presented to them, so that in fact the number of ants coming out to follow the trail was also increased. This shows then that individual scouts are able to affect the number of nest mates recruited with the quantity of trail substance they release, by adjusting this to the quality of the food source. The chemical composition of the trail substance is known neither in the case of *Acanthomyops interjectus* nor in that of the fire ant. We do know from the investigations of Barlin *et al.* (1976), however, that *Solenopsis richteri* and *Solenopsis invicta* lay species-specific trails, whilst *Solenopsis geminata* and *Solenopsis xyloni* probably produce the same trail substance. The trail substance of *Solenopsis richteri* consists of a number of components, the main ones having a molecular weight of 218, the summation formula being $C_{16}H_{26}$. The smallest quantity of this substance which triggers a following response amounts to 10 fg or 10^{-14} g per cm. How small this quantity is, becomes clear perhaps from the fact that 1 g of this substance would suffice for a trail one thousand million km in length, 55 times the distance between the earth and the sun.

One question which has been much disputed is that of the directional specificity of trails. According to Bethe (1898), the trails of ants are polarized, so that the animals are able to recognize from them their way to the feeding place or the nest, as the case may be. However, this hypothesis has since been refuted, both for ants (Carthy, 1951b; Wilson, 1962c; Leuthold, 1975) and termites (Tschinkel and Close, 1973). According to Leuthold, for example, inexperienced *Crematogaster* workers, coming across a trail of their own species, run in equal numbers in both directions of the trail. Even under natural conditions, one can occasionally observe errors, such as when workers carrying food travel in the trail direction leading away from the nest.

Among leaf-cutter ants, who give orientation aid with their trail substance as well as recruiting, the trail pheromone of one of these ants has been successfully analysed for the first time (Tumlinson *et al.*, 1971, 1972). In the case of *Atta texana*, the trail substance consists mainly of methyl 4-methylpyrrol-2-carboxylate (Fig. 5.9), a substance which has since also been synthetically produced. The behavioural response threshold lies at 0.8 pg per cm, or 0.8×10^{-12} g, or 3.48×10^{3} molecules per cm. We have since discovered that this substance is also produced as a trail pheromone by other leaf-cutter ants (*Atta cephalotes*; Riley *et al.*, 1974) and that it produces a behavioural response in further leaf-cutter species, e.g. *Atta colombica, Atta laevigata, Acromyrmex octospinosus, Acromyrmex versicolor, Trachymyrmex septentrionalis, Trachymyrmex urichi, Apterostigma collare* and *Cyphomyrmex rimosus,* but not in the case of *Atta sexdens* (Riley *et al.,* 1974; Robinson *et al.,* 1974). If the ants are presented with similarly constructed molecules as a trail, as has been done by Sonnet and Moser (1972, 1973), it can then be demonstrated that the most effective substances are those which have a free NH group with a related ester function and a substituent at ring position 4. Whilst the steric proportions of the carbonyl group are

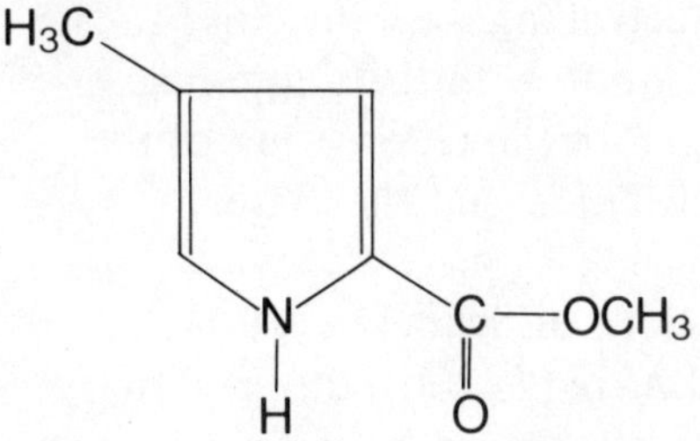

Fig. 5.9   Methyl-4-methylpyrrol-2-carboxylate.

crucial for behavioural effectiveness, one can on the other hand modify them with nitrogen and carbon, carried by the methyl group, without loss of effectiveness.

Trail substances have also been proved to exist in other subfamilies including the Cerapachyines and Dorylines, among certain Pseudomyrmecine species and in exceptional cases also among Ponerines. The Myrmeciines seem by contrast to lack trail substances entirely.

The Dolichoderines produce their trail substance in a special gland quite independent of the defence apparatus, the Pavan's gland (Fig. 5.6). The excretory duct opens to the outside of the body between the fourth and fifth segments of the rear section of the body (Wilson and Pavan, 1959). For some time, the main species investigated were those of the *Iridomyrmex, Tapinoma* and *Monacis* genera. The striking fact here is that in contrast to species of *Lasius* and many leaf-cutter ants, the trail substances of the Dolichoderines investigated so far are markedly species-specific.

Among the Dorylines, the army ants, the armies orient themselves using trail pheromones, which are stored in the end of the intestine as with the Formicines, and released via the anus. The cohesion of the colony during travelling also depends, apart from mechanical contact, on the effect of the trail pheromones. If the trail once becomes broken, the portion of the army following stops, and requires a certain amount of time to rejoin the rest. The trail itself is laid by a group of workers, who march at the head of the army, separated from the rest by a few centimetres. Trail laying does not involve a particular group of workers, but 'temporary pioneers'. Every worker of a colony treading new ground lays a trail; this is so with the worker group at the head of an army, who then withdraw and are relieved by other workers. The workers new to the head of the army creep along, keeping close to the ground and thoroughly feeling the substratum with their antennae. Each of them marks the ground beneath with the rear section of its body, at the same time releasing trail substance, and then later withdraws (Schneirla, 1971).

As far as we know from the investigations of Schneirla (1944), Rettenmeyer (1963) and Torgerson and Akre (1970b), the specificity of Doryline trail substances is quite minimal. As a rule, various species of the same genus follow their trails at different times, and in many cases also follow the trails of species from other genera. Apart from this, the trails of army ants can also be picked up by beetles, who follow these ants. Examples of this are the Staphylinid and Histerid species *Euxenista* and *Vatesus* respectively (Akre and Torgerson, 1969; Torgerson and Akre, 1970a) or also the Carabid species *Helluomorphoides texanus,* which follows the trails of *Neivamyrmex nigrescens.* It feeds mainly on the brood of the ants, and is itself

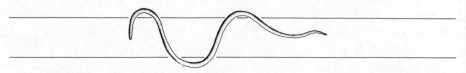

Fig. 5.10   The snake *Leptotyphlops dulcis* following the artificially laid trail of an army ant (after Watkins *et al.*, 1967, adapted).

protected from being attacked by ants by means of a defence substance, which frightens them off (Plsek *et al.,* 1969).

Interestingly, in addition to these species, there are also vertebrates who follow the trails of army ants. The snake *Leptotyphlops dulcis* was found during the nocturnal travels of the army ant species *Neivamyrmex nigrescens.* It was proved in the laboratory that this blind snake in fact follows the chemical trail of the army. Here, artificial trails were laid with a mixture of the contents of the rectal bladder of the species *Neivamyrmex nigrescens* and distilled water, and it was possible to observe that the snakes followed these trails (Fig. 5.10). In the natural environment, the snakes follow the army ant trails and feed on ant broods (Watkins *et al.*, 1967). The snakes are chiefly protected from the ants by substances from their 'cloacal sack'. This was initially indicated by the observation that the cloacal sack is empty after the snake has been attacked by ants (Gehlbach *et al.*, 1968). It was then determined that the secretion of the cloacal sack acts as a chemical deterrent (Watkins *et al.*, 1969), and the chemical analysis of this substance finally revealed that it consists of two main components, glycoprotein and fatty acids, with which the snakes coat the whole of their bodies. The glycoproteins probably have the effect of rendering the snakes slippery and difficult for the ants to grip, whilst the fatty acids frighten the ants away (Blum *et al.*, 1971).

Ants are also able to deposit deterrent materials, or repellant substances, during the acquisition of food. It has long been known in the case of the robber ant *Solenopsis fugax,* that they situate their nests in the immediate vicinity of other larger ant species. The nests consist of numerous nest chambers, connected by a fine mesh of narrow passages. These passages also extend into the nests of the alien ant species, and are used by the robber species to succeed in reaching the alien brood, which they then remove. Since ants usually keep a close watch on their own brood, the question was initially asked as to how the robber ants manage to hold their own in the alien brood chambers. We have since discovered that *Solenopsis fugax,* like *Monomorium pharaonis,* which probably also lived as a robber ant in its native East Indies, not only marks its passages with recruitment pheromone from the Dufour gland, but also with repellant substance from its poison gland. *Monomorium* disposes of feeding rivals at common feeding places with this substance, and *Solenopsis* succeeds in making its way unhindered into the alien brood chambers. How similar these chemical strategies in fact are with *Solenopsis* and *Monomorium* emerges from the fact that each is unable to distinguish the trail of the other species from its own, and also from the fact that neither reacts to the repellant substance of the other species (Hölldobler, 1973c).

5.4 Communication during nuptial flight and pairing

Apart from relatively few exceptions, the sexual animals of ants have wings. They mostly leave the nest shortly after hatching and commence their 'nuptial flight'. Such a nuptial flight of the young sexual animals usually brings together not only those of one colony, but also those of different colonies, who pair either on the wing, or, as is more frequently the case, immediately after the flight. The freshly inseminated females then set about founding a new colony if they are not taken in by other colonies, whilst the males, who have already fulfilled their life purpose, usually die shortly after pairing.

During the nuptial flight and pairing, the behaviour of the young sexual animals needs to be well synchronized. Males and females, for example, must take flight at the same times in order to come together on the wing. It is also important that the swarm times of the sexual animals of different colonies should be synchronized, unless it is simply a question of inbreeding. Furthermore, the flight directions of the flying ants must be matched in such a way that the swarming sexual animals of different colonies actually do meet. Finally, for a successful insemination to take place, males and females must be able to recognize each other and be induced to copulate.

We are still a long way from knowing all the factors which trigger a swarm to take flight and affect its further progression. Also there are — as is only to be expected — considerable differences between different ant species. We are, however, already able to distinguish two different groups of factors. Firstly, there are environmental factors, such as temperature and time of day, which essentially determine and synchronize the point at which the swarm will take to flight; then there are particular landmarks, such as trees, hills, towers and even lakes (Sanders, 1972), which guide the sexual animals of different colonies towards each other. The effect which these environmental factors can bring about may be seen by the fact that in some cases the fire brigade has been called out because the dense clouds of flying ants, making for a building being used as a marker, have been mistaken for clouds of smoke. In addition to environmental factors, special pheromones have been discovered, which similarly contribute to the synchronization of the behaviour of the sexual animals. Hölldobler and Maschwitz (1965) have investigated the swarm behaviour of the carpenter ant *Camponotus herculeanus,* which nests in trees, in detail. They describe the 'taking off' of a swarm in the following manner: 'A few males begin with short flights, of distances between 2 and 4 m; they land at this point and usually walk back to the nest. Shortly afterwards, the males crawl higher up the shaft leading to the colony exit, until they have climbed about 2 m, whilst a continuous stream of further males pours out of the opening. The circulatory movement from the nest towards the light and back to the nest, which could be observed during the sun reaction, no longer takes place. All the males tend to move towards the light and place themselves almost perpendicularly to the shaft, until the take-off eventually takes place like a swelling avalanche. Whilst males inside the nest move according to negative phototaxis, at the time of the swarm they display a positive phototactic response. The females, who have since gathered in greater numbers at the flight exit, exit in waves by the hundred, shortly before the high point of the males' take-off, and walk in wide columns, usually to the uppermost reaches of the tree. From here they take to flight. The males swarm upwards at quite a steep angle; on the other hand, the heavier females gain height quite slowly in unaided flight. A joining together of separate swarms is never observed: males and females scatter diffusely in the open air.'

The particularly striking feature in this area is that the take-off frequencies of the males gradually increase, and that the females begin at the precise point at which the males are almost at the high point of swarming.

It had already been observed in the laboratory that carpenter ant females react strongly to a male secretion which the males produce in their mandible glands, so Hölldobler and Maschwitz (1965) were able to test whether this secretion plays any part in swarming behaviour. They did in fact find that the females can be induced to swarm with the mandible gland secretion of the males, as long as the temperature and time of day were also favourable. They also discovered that the males which were about to take off had completely full mandible glands, whilst the glands of the males who had just taken off were empty. Thus the conclusion was drawn that the males, shortly before taking to flight, odorize the area surrounding the swarm exit with the secretion of their mandible glands. The strongly smelling secretion has no effect on other males, but attracts the females to leave the inside of the nest and to take off just after the majority of males have done so. During the nuptial flight itself, the mandible gland secretion has no effect on the females. Both sexes orient themselves optically at this time by flying towards marked points such as the high tops of trees, and mate there.

The secretion of the mandible glands of carpenter ants was chemically investigated by Falke (1968). He found five different substances in the secretion, of which he was able to identify four; however, these were not the behaviourally effective components. The mandible gland secretion also plays an important part in swarming among *Lasius* species, as emerges from the research of Law *et al.* (1965) and of Hölldobler and Maschwitz (Hölldobler, 1973d) with various *Lasius* species.

Female sexual attraction substances were proved to exist in other ant species, for example the *Formica* species. *Formica* species differ from the carpenter ants in many respects since frequently in one colony, only males or only females are formed, not both sexes together. This applies to the species *Formica montana* and *Formica pergandei*, whose swarming behaviour has been more closely researched by Kannowski and Johnson (1969). The males, which in this case start from different nest mounds from the females, firstly climb up plants in the vicinity of their nest and remain there motionless. If a freshly hatched female then appears within a radius of 15–20 m, the males become agitated at their position on the plants, and finally fly directly to the female and inseminate her. The inseminated females then rid themselves of their wings and seek suitable places to found a new colony. The males on the other hand seek further females who have not yet been inseminated. Freshly hatched males who after a 10-30 min. wait have not yet found a female, change over to conducting search flights, usually flying close to the ground. The freshly hatched females, having left their nest are, however, only rarely observed in flight. They favour going on foot, climbing plants in the vicinity of their nest, where they remain motionless. From this position they attract the males of their species by means of odour. This emerges from research conducted by Kannowski and Johnson in the laboratory. They were unable, however, to discover in which gland the females produce their sexual attraction substance.

Talbot (1959, 1971, 1972) discovered similar nuptial flights in the species *Formica obscuripes* and *Formica dakotensis* as Kannowski and Johnson had done in the cases of *Formica montana* and *Formica pergandei*. Here also the females position themselves on grass and other plants in the vicinity of their nests, whilst the males search for females ready for insemination in flights close to the ground. The males in these species also are very probably attracted by female pheromones.

It has been demonstrated that among ant species of the Myrmicine subfamily, a sexual attraction substance is produced in the poison gland. The Myrmicine *Xenomyrmex floridanus* is very widely found in the mangrove swamps of Florida. Very little is know about their nuptial flight in the natural environment, but it is well known from the laboratory experiments of Hölldobler (1971a), that the winged females of this species attract their males very effectively. The attraction effect on the males remains the same when one presents them with the squashed abdomen of the females or only the secretion of their poison gland instead of the intact winged females. This demonstrates that the *Floridanus* females produce a sexual attraction substance in their poison gland which, as emerges from further experiments by Hölldobler, not only attracts the males but also stimulates them sexually.

Among the socially parasitic ants, there are a few species whose sexual forms have no wings. Thus for example, the species *Harpagoxenus sublaevis* already mentioned, has only a few winged females. These, and a number of other species are, of course, unable to undertake a nuptial flight. Buschinger has investigated the mating behaviour of these species and discovered that here too a sexual attraction substance plays a part.

Harpagoxenus sublaevis was previously regarded as a rather rare species, until Buschinger (1966a,b) found them to be relatively abundant in a number of places and was able to breed them in the laboratory. Buschinger (1968a,b) was able to observe that the females ready to be inseminated left the nest between 18 and 22 h under laboratory conditions, and climbed raised areas in the vicinity. The animals generally remain motionless on these raised areas, with the rear section of their bodies pointing upwards, having opened their cloaca and outstretched the sting. A male which is ready to mate approaching the female, moves directly towards her and inseminates her. Copulation lasts altogether between 15 and 20 s. Afterwards the male drops behind the female (*see* Fig. 5.11). Upon a very male gesture on the part of the female, a bite in the rear section of the body, the male breaks the mating link. Marikovsky (1961) found that in the case of the wood ant *Formica rufa* in western Siberia, females can on this occasion, even tear off the rear section of the male's body.

The behaviour of the *Harpagoxenus* females before mating is reminiscent of ventilating bees, and has as a result of this been described by Buschinger as 'attraction ventilation'. Buschinger found attraction ventilation also among the similarly socially parasitic ants *Leptothorax kutteri,* as well as among *Harpagoxenus* (1971a), and in addition among *Leptothorax gösswaldi* (1974), *Doronomyrmex pacis* (1971b; Fig. 5.11), among the non-parasitic guest ant *Formicoxenus nitidulus* (1976b) and the completely independent species *Leptothorax muscorum* and *Leptothorax gredleri* (Buschinger, pers. inf.).

Buschinger (1972a) was able to show in laboratory experiments that the *Harpagoxenus* males fly to ventilating females as well as prepared

Fig. 5.11   Mating of *Doronomyrmex pacis*. The male drops behind at the end of mating and is bitten in the rear section of the body shortly afterwards by the female, whereby the mating linkage is broken. The male is able to inseminate further females afterwards, such as the two ventilating females, which can also be seen apart from the mating pair (Photo — A. Buschinger).

poison glands from a distance of 3-4 m. Usually, however, they do not land directly on the odour source, but within a radius of a few centimetres and begin an intensive search. If they meet the females during this, then copulation follows immediately. If they only find poison gland secretion, however, the males do not try, like the *Xenomyrmex floridanus* males, to copulate with the odour source. In the case of *Harpagoxenus,* direct contact with the female is necessary for the stimulation of male copulation behaviour, during which contact a surface pheromone is active on the body of the female. Thus, if the body of the female ready for mating is washed with liquid benzine, she loses her attraction for the male. Females thus treated are unable to stimulate any further copulation behaviour (Buschinger, 1973b).

Buschinger also investigated the effectiveness of sexual attraction substances between socially parasitic species, which occur only rarely in nature. *Doronomyrmex* was first described by Kutter in 1945, and *Leptothorax gösswaldi* in 1967, whilst *Leptothorax kutteri* was first discovered by Buschinger in 1965. The sexual attraction substances of

Leptothorax kutteri and *Doronomyrmex pacis* are mutually effective, which is astonishing, since sexual attraction substances are usually strictly species-specific. In Buschinger's experiments, the *kutteri* males were attracted by the *Doronomyrmex* females and tried to inseminate them, failing for mechanical reasons. The *Doronomyrmex* males were also attracted by the *kutteri* females, and even mated successfully. In this manner, Buschinger (1972b) obtained seven hybrid females. In natural conditions, such a cross would probably be ruled out by the fact that the females of the two species ventilate at different times of day. In Buschinger's experiments, the crosses were only successfully made by heating the colonies of the different species at different times of day from the low night temperature to between 24 and 27°C, so that the ventilation times were synchronized. *Leptothorax gösswaldi* males also react to the sexual attraction substances of other species, specifically *Doronomyrmex pacis, Leptothorax kutteri* and *Harpagoxenus sublaevis.* Conversely, the sexual attraction substance of *Leptothorax gösswaldi* is only effective for *Doronomyrmex pacis* males and the males of *Leptothorax kutteri* (although in the latter case only to a very small degree), but has no effect on the males of *Harpagoxenus sublaevis.* Copulation occurred between *Leptothorax gösswaldi* males and *Leptothorax kutteri* females, and between *Doronomyrmex pacis* males and *Leptothorax* females, but as a result, young were produced in only the first case (Buschinger, 1974).

In the case of *Monomorium pharaonis,* we also know that the females produce a sexual pheromone, which is given off by the animals ready to mate, and attracts and sexually stimulates the males. As distinct from the Myrmicines mentioned above, however, the sexual attraction substance is here produced not in the poison gland, but in the Dufour gland and the Bursa side sack. Workers have no sexual pheromone in these glands, and even the contents of the glands of older females which have already laid eggs in the nest are neither attractive nor sexually stimulating for males (Hölldobler and Wüst, 1973). We know from the investigations of Buschinger and Peterson (1971) and Peterson and Buschinger (1971b) that the sexual animals copulate in the nest or in the close vicinity of the nest, actually only a few days after hatching. The sexual attraction substance thus does not need to be effective over long distances. It has a maximum reach of only 6 cm and has an attraction effect from this distance. It has a sexually stimulating effect on the males only in higher concentrations, at a distance of 2-3 cm from the young females. For successful copulation to take place, further specific behaviour patterns are required on the part of the female ready to mate, as well as the male, including touching with the feelers and the offering of the rear section of the female body.

Apart from some representatives of the Formicines and Myrmicines, one can also recognize indications of sexual attraction substances among the Pseudomyrmecine subfamily. Janzen (1973) observed the progress of copulation among the obligatory Acacia ant species *Pseudomyrmex venefica,* which lives exclusively in the hollow thorns of acacia trees. The young females of this species climb out of the same branches in which the males hatched a few hours earlier. Whilst the males then unfold their wings and leave the tree where they hatched with a summary flight, the females walk only a few centimetres, climb on to a leaf or a thorn and remain in one position, reminiscent of the 'attraction ventilation' of the species investigated by Buschinger. They stay in their places with the rear section of their bodies raised, the cloaca opened wide. Males coming from foreign nests, landing on the tree and meeting with these females, are immediately induced to copulate. Evidence of attraction ventilation has also been described in the case of the Ponerines, such as the wingless females of *Rhytidoponera* species which live in Australia and are close to the forbears of the Myrmicines (Haskins and Wheldon, 1965), the winged females of *Amblyopone pallipes* found in North America, and the South American *Dinoponera grandis* (Haskins, according to pers. inf. from Buschinger). These last examples are particularly noteworthy: they permit us to conclude that there is a possibility of attraction ventilation representing an ancient heritage of the forbears of the Ponerines (Buschinger, *pers. inf.*).

5.5 Communication during exchange of food

The exchange of food among individuals of a colony is of unusually great significance among ants. During the exchange of food, the liquid food carried in the crop is divided among the members of the colony, who thus relinquish the search for their own food and are able to devote the whole day to other tasks inside the nest. The exchange of food is, therefore, an important precondition for the division of labour between those animals serving inside, and those serving outside the nest, a feature which is strongly developed to a greater or lesser degree in almost all ant species. In addition, glandular produces may also be given during the exchange of food, which are important, for example, for caste differentiation.

Apart from liquid form, food may also be handed on in solid form, in the shape of the so-called 'trophic eggs'. These eggs are unable to develop and are fed mainly to the queens and the larvae, but in rare cases also to workers. These features are the case in the colonies of *Iridomyrmex humilis* (Torossian, 1961), *Pogonomyrmex badius* (Wilson, 1971), *Dolichoderus quadripunctatus* (Torossian, 1959) and *Plagiolepis pygmea* (Passera, 1965).

Among *Myrmecia gulosa* (Freeland, 1958) and *Dolichoderus quadripunctatus* (Torossian, 1959), there are special begging signals for the trophic eggs: the worker places her mandibles on to the mouthparts of the nest mate to whom she is begging and strokes the tips of the nest mate's mandibles with her palpae.

In the majority of cases, the larvae receive either liquid food, or are presented with whole, captured animals. Among Myrmeciines, the ancient Ponerine *Amblyopone* (Haskins and Haskins, 1951) but also among *Myrmica* and *Aphaenogaster* (Le Masne, 1953; Buschinger, 1973a), the larvae are placed on whole, captured animals, or large pieces of them (Fig. 5.12). We shall see later how this method of feeding has been interpreted by Malyshev (1966) as an indication of tree-dwelling forbears of the ants.

The giving of liquid food (trophallaxis) occurs among almost all species of ants, with the exception of only a majority of Myrmeciine species, a few Ponerines, such as *Amblyopone,* and particularly specialized Myrmicines such as harvester or leaf-cutter ants. The ants' crop acquires a new significance in the adaptation for trophallaxis: it serves not only as a food store for the individual animal, but also, together with the crops of all the other nest mates, as a reserve store for the entire colony ('social stomach').

Among certain species, which have to withstand long periods of drought, there are special storing animals, which collect more food than other nest mates in their crop. With a filled crop, the rear section of the body of this 'replete' ant (Fig. 5.13), is swollen to such an extent that the animal is no longer able to move, or only with great difficulty. They then hang, as in the case of *Myrmecocystus horti-deorum,* like living honey pots in the nest chambers and are tapped according to the requirements of the nest mates. In the case of *Myrmecocystus mexicanus,* Burgett and Young (1974) have investigated in detail the form in which the food is stored. In the stored food of the replete ant they found fats (glycerolester and cholesterolester), as well as carbohydrates (glucose, fructose and maltose).

In the older queens of various species, a second crop is situated in the

Fig. 5.12   Aphaenogaster subterranea workers carrying the larvae to a captured animal and placing them on it (Photo — A. Buschinger).

Fig. 5.13   Replete of the honey ant *Myrmecocystus melliger,* which hangs with a full crop from the roof of the nest, and gives its stored food to nest mates when required (after Wheeler, 1923, adapted).

thorax of the animals, very probably taking up the space which would have contained the wing muscles in younger queens. We have been unable to discover the function which is served by this thoracic crop – whether it merely increases the storing capacity of the abdominal crop, or whether it also has an additional function (Petersen-Braun and Buschinger, 1975).

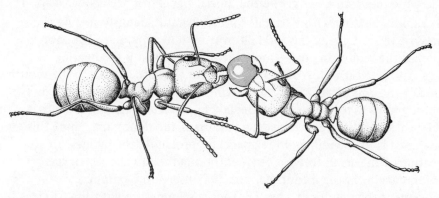

Fig. 5.14   The giving of liquid food (trophallaxis) between two *Formica sanguinea* workers (after Hölldobler, 1973a, adapted).

When the contents of the crop are being given, the food is first regurgitated, and then brought on to the outstretched labium, from where it can be taken by the ant being fed (Fig. 5.14). This feeding behaviour may be initiated by the receiver as well as the giver.

Among *Formica sanguinea,* for example, the initiative for reciprocal feeding occurs much more frequently in the receiver than the giver (Wallis, 1961; Hölldobler, 1973a). One finds by contrast, that behaviour patterns offering food are favoured by those workers returning to the nest with a full crop. Hölldobler describes this behaviour in the following manner: 'They walk towards nest mates with open mandibles and outstretched labium, orienting themselves by touch with the antennae. Mouth to mouth contact, or a light touch of the labium is sufficient to stimulate regurgitation in the food carrier. If the ant offering food finds itself without a receiver for any length of time, it often regurgitates a small amount of food without tactile stimulation, and carries this around in the meantime between its mandibles, until it is finally able to divest itself of the food on the ground or the wall of the nest. Usually, however, the honey drop is taken by a nest mate before this happens, which can then lead to a more protracted feeding.'

The behaviour of the ant begging for food, as represented in the analysis of normal (24 frames/s) and high frequency film (200-450 frames/s) of the species *Formica sanguinea,* are described by Hölldobler (1973a) in the following way: 'The ant requesting food first rapidly runs its antennae in a trilling motion over a nest mate, which thereupon usually turns towards the requesting ant. The animals then stand face to face, the requesting ant giving the 'initial signal' for the giving of food by rapidly beating with its small forelegs, simultaneously feeling the head of the ant giving the food all over with its antennae. The more intense the requesting behaviour, the more markedly the beating of the legs is directed inwards, and the more sure they are of hitting the mouth area of the animal at which the begging behaviour is being directed. The photographs taken from the underside in particular clearly show that the beating is directed against the labium of the donor ant. Since this ant spreads its mandibles wide apart and protrudes the labium, the beating of the small forelegs frequently hits the target and triggers regurgitation. In the meantime, one can observe the pumping movements of the donor's gaster. Tactile stimulation with the forelegs is maintained during the actual flow of food as a rule, but the receiver also touches the donor's labium in the 'trilling' motion with its maxillae and feels the donor's head with its antennae; the donor for its part keeps its antennae wide apart. Only when the readiness to give food begins to abate does the donor ant place its antennae at a sharper angle and raise its front extremities more and more. The beating with the front legs by the begging ant immediately becomes more intense, so that a second regurgitation phase may be triggered. Finally, the donor ant raises its forelegs, and may even begin to strike with them, whereupon feeding usually discontinues and the animals separate.'

Hölldobler (1973a) was able to imitate successfully the mechanical signals of begging workers among various *Myrmica* and *Formica* species. When he stimulated workers returning to the nest with a full crop, stimulating them mechanically on the labium, they regurgitated the contents of the crop. Hungry larvae usually beg for food in much the same manner. The *sanguinea* larvae, for example, beg by bending the front part of the body up and down, swinging it this way and that, and by nudging the worker's labium. According to Heyde (1924) and Hölldobler (1966), freshly hatched workers and males do not beg for food spontaneously; they are initially offered food intensively by older ants and only later begin to beg for themselves.

Apart from oral trophallaxis, there is also, for example in *Zacryptocerus varians,* an abdominal form of trophallaxis, whereby liquid food is given via the anus. Begging and offering signals were also found in this species. During begging, the worker feels the tip of a prospective donor's rear segment with its mouth parts, and thoroughly licks the area. When food is being offered, the workers raise their hind parts and expose the anal zone as much as possible (Wilson, 1976).

5.6 Recognition of nest mates

Ants are usually able to distinguish the members of their own colony from those of other colonies. One can see this, for example, from the fact that animals from an alien colony are usually violently attacked and killed. One exception from this rule is that of freshly hatched young workers: in contrast with their older siblings, they are not attacked in alien nests of their own or closely related species, probably because these young workers produce a substance which inhibits aggression in ants of their own or closely related species, as Jaisson (1972) discovered with *Myrmica rubra* and *Formica polyctena.*

The phenomenon of the incompatibility between individuals of different colonies was already known and investigated early on by Forel (1874), Bethe (1898), Fielde (1901, 1905), Piéron (1906), Brun (1913) and other researchers. They reached the conclusion that ants evidently recognize one another by their sense of smell. Regarding this, we still do not know the exact origin of the scent and which individual substances are involved. We speak about a 'colony scent' without being able to be very exact. In recent times, investigations of this subject have been conducted mainly by Lange (1958, 1960a,b), Soulié (1960b) and by Hangartner *et al.* (1970).

According to the results of Brun (1913), there is a species-specific hereditary component in the scent of the colony and a 'local component', which has also been confirmed by other authors. The local component is evidently as much affected by the food on which the colony feeds as by the composition of the nest material, as emerges, for example, from Lange's investigations. Lange divided a laboratory *Formica polyctena* colony into different groups and kept these groups separated. He proved in regulated tests the extent to which they recognized one another as members of the original single colony. The reciprocal exchange of food served Lange as a gauge of the recognition of the members. It was finally shown that workers of the different groups still recognized each other as members of the same colony after months of separation, provided they had been kept under identical conditions. If however, the different groups had received different food or nest materials, then the exchange of food became restricted after only a few days, and enmity very soon prevailed among the groups. This shows that odour substances from the substratum of the nest, a 'nest scent', evidently also affects the colony scent. The ability of ants to sense this nest scent emerges from the research of Soulié (1960b) with *Crematogaster,* and of Hangartner *et al.* (1970) with *Pogonomyrmex badius.* In selective experiments, for example, Hangartner and his colleagues found that the American harvester ant, *Pogonomyrmex,* unequivocally favoured the nest material of its own nest over substrata from other nests and over substrata with identical physical properties (moisture, size of particles). Obviously, however, not all individuals of an ant colony have an identical scent, or it would not be possible, for example, for fire ant workers to distinguish their queen from their fellow workers, as Jouvenaz *et al.* (1974) have shown. The queens of these species (*Solenopsis invicta* and *Solenopsis geminata*) must possess a special odour note which sets them apart from their workers. The workers of both species are attracted not only by their females, but also by small pieces of filter paper which have been soaked with an organic stimulating medium (hexane), and with which the queens had previously been washed. Jouvenaz and his colleagues were able to show in the case of *Solenopsis invicta* that filter papers treated with extract from their own queen were more attractive for the workers than those treated with extracts from queens of the same species but alien colonies.

Watkins and Cole (1966) have already made similar observations with six different species of South American army ant, which were attracted by pieces of filter paper which had been in contact with the queens for a while, as well as by the queens themselves. In these species also − of the genera *Neivamyrmex* and *Labidus* − it could be shown that a filter paper which had been in contact with the ants' own queen was more attractive to the workers of a colony than other papers which had been coated with the odour substances of queens of the same species but from alien colonies.

However it is not only the queens, but also the brood of a colony which has the benefit of attractive odour substances, by means of which they can be recognized by the workers in the darkness of the nest. One is easily convinced of the capability of the larvae to attract the workers if one removes an ant nest from the natural environment into an artificial nest: the workers are then above all preoccupied with collecting together their scattered larvae, pupae and eggs and gathering them into a pile.

Brian (1975) found with *Myrmica rubra* and *Myrmica scabrinodis* that the larvae are recognized by a pheromone which covers the whole surface of their bodies. The chemical composition of this pheromone is still unknown; so far we know only that it is very slightly volatile and that it is soluble neither in hexane nor in water, methanol or 70% ethanol, but that it may be washed off with acetone, ether and chloroform. Apart from this, it emerges from investigations that this chemical recognition signal for larvae is not species-specific, at least not between *Myrmica rubra* and *Myrmica scabrinodis*. In this case, each species handles and cares for the larvae of the other species as if they were their own brood. The pupae and possibly also the eggs are recognized by *Myrmica* workers by means of other chemical signals, whilst the various stages of the larvae's development are probably distinguished by means of the number of their bristles.

Chemical recognition signals have also been demonstrated to exist for the larvae of other ant species, the fire ant *Solenopsis saevissima* and *Solenopsis invicta* (Glancey *et al.,* 1970; Walsh and Tschinkel, 1974) and the leaf-cutter ant *Atta cephalotes* (Robinson and Cherrett, 1974). Among fire ants also, the larvae and pupae are recognized by their workers by means of chemical characteristics, and here also the pheromones are distributed over the entire surface of the brood. In contrast with *Myrmica,* however, the pheromone of the *Solenopsis* larvae is soluble in hexane and in methanol; apart from this, however, the role played by mechanical signals among fire ants for distinguishing the various stages of larva development, if it exists at all, is only very small. The brood of the leaf-cutter ant *Atta cephalotes* is also recognized by means of chemical signals. If the pupae of this species are killed by being exposed to cold then, in contrast with the larvae, they lose none of the chemical attraction for the workers, but they do if they are washed off with hexane.

The chemical signals of the brood are not recognized by the workers from 'birth' onwards, but are learned during a sensitive phase shortly after hatching, probably irreversibly. Such a learning process is described as imprinting and has been demonstrated in several ant species by Jaisson (1975). He separated young *Formica polyctena* workers from the mother colony on the day of hatching and placed them in groups of

250-300 individuals. He gave pupae of various species into the care of these workers, such as the pupae of *Formica sanguinea, Formica nigricans, Camponotus vagus* and *Lasius niger,* and was able to observe that these pupae were willingly cared for by the workers. Those workers, which during the first 15 days after hatching had cared for the pupae of other species, subsequently accepted further pupae to care for only from the previously named species, whilst the pupae of other species, including their own, were treated as food and eaten. Groups of freshly hatched workers which had received no brood were later incapable of caring at all for pupae given over to them.

Among ants, dead as well as living nest mates are recognized by means of chemical signals. On the basis of these signals, the corpses of workers are gathered together and transported out of the nest, which is described as 'necrophoric behaviour'. The chemical stimuli for necrophoric behaviour have been demonstrated to exist among the American harvester ant *Pogonomyrmex badius* (Wilson *et al.,* 1958), the fire ant *Solenopsis saevissima* and *Solenopsis invicta* (Blum, 1970; Howard and Tschinkel, 1976) and among the Australian species *Myrmecia vindex* (Haskins and Haskins, 1974). It has emerged from various experiments that these stimuli actually do involve chemical signals. Howard and Tschinkel treated 'effective' corpses with stimulus medium and then observed that they lost their ability to trigger necrophoric behaviour. Wilson and his colleagues soaked pieces of filter paper with extracts from dead *Pogonomyrmex* workers and found that these pieces of filter paper were treated as dead nest mates and placed in the 'cemetery'. The extracts from dead fire ants were analysed by Blum; he found myristol, palmitol, oil and linol acids and other fatty acids, which were able to trigger necrophoric behaviour separately as well as together. These substances are byproducts of the decomposition of triglycerides, arising from bacterial decomposition. Bacteria which are able to hydrolyse triglycerides were found in large quantities in ant corpses. This hypothesis, that the substances arising from bacterial decomposition stimulate the necrophoric behaviour, is contradicted, however, by the experiments of Howard and Tschinkel, according to whom the chemical stimulus of freshly killed fire ants is missing, but soon develops and becomes most powerful after one hour. The stimulus effect is equal whether the animals have been killed by heat or by cold, which indicates that these signals are not the product of an enzymatic process. The signals cannot be sensed by the workers through their sense of smell, only by means of direct contact. Howard and Tschinkel concluded from their experiments that chemicals triggering necrophoric behaviour are concealed by other pheromones whilst the ants are alive, and which themselves decompose after death, in fact quicker than the chemical triggers for necrophoric behaviour. These chemical stimuli are not identical amongst the different ant species. This emerges from experiments conducted with *Pogonomyrmex badius* (Wilson *et al.,* 1958) and *Myrmecia vindex* (Haskins and Haskins, 1974). According to these experiments, triethanolamine is the most effective stimulus for necrophoric behaviour for *Myrmecia vindex,* whilst it has no effect at all with   *Pogonomyrmex badius.*

Mühlenberg and Maschwitz (1976) observed in a species of bug found in Ceylon (*Acanthaspis concinnula*), how they load themselves on to the ants' backs, capture them and suck them dry. They are held in their positions by means of bristles and hairs and additional spun threads. The meaning of this quite remarkably demanding behaviour for the bugs cannot rest merely in protection against enemies who would feed on them, as is the case, for example with Cassidide larvae, having been demonstrated by Eisner *et al.* (1967); the larvae could also procure advantages from the exploitation of ants, since the ant corpses would mask their scent for the predators. It could also be the case that the predators themselves take advantage of the necrophoric behaviour of the ants, thereby reaching their prey more easily (Mühlenberg and Maschwitz, 1976).

6 Caste differentiations

6.1   The various ant castes

One of the most decisive steps in the development of the social insects was the formation of a special caste of workers, which, as in the case of bees and wasps as well as ants, was developed in the female sex alone, not the male. Among some species from the Ponerine subfamily, for example, *Hypoponera eduardi* (Fig. 6.1), it is possible to distinguish

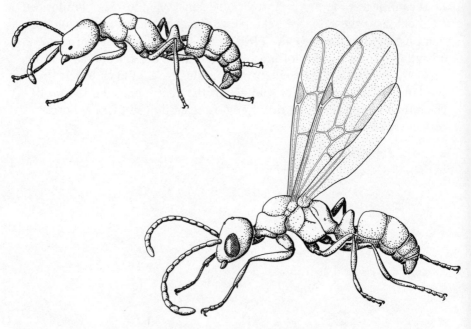

Fig. 6.1   Two different forms of male (with and without wings) of the species *Hypoponera eduardi* (after Le Masne, 1956a, adapted).

different male forms (those with and without wings), but this is not sufficient to permit us to speak of different castes in this instance. The different exterior appearance of these males is not connected with any difference in function within the colony. Apart from this, no-one has been able to observe the occurrence of both winged and wingless males in the same colony (Le Masne, 1956a).

Males remain in the colony for varying lengths of time; in the case of the wood ant *Formica polyctena,* the males leave the nest as early as a few days after hatching, whilst carpenter ant males (*Camponotus herculeanus* and *Camponotus ligniperda*) remain throughout the winter as fully grown animals and only start on their nuptial flight the following year. During the time they remain in the nest, the males are fed and cared for by workers. This does not mean, however, that they block the social flow of food, as was previously assumed (*see* Escherich, 1917; Goetsch, 1940). On the contrary, they pass a proportion of the food they receive on to other nest mates, but do not feed any larvae (Gösswald and Kloft, 1960a; Hölldobler, 1966).

Among the females in some species, there are, apart from the queen and worker castes, further sub-castes, such as the replete ants of the honey ant species already mentioned, soldier ants and the major and minor ants. Wilson (1971) on the other hand, not only specifies the males as being a caste in their own right, but further distinguishes between the ergatoid males (without wings and ressembling workers), the queens, workers, soldiers, ergatogynes (an intermediate form between worker and queen) and dichthadiiforme ergatogynes (a special form of wingless female among Dorylines) as castes in their own right.

6.2 Polyethism

Ants undertake — according to their caste — various tasks in the colony. If one compares these caste-specific tasks, it strikes one that the burden of work is unevenly distributed. Apart, at most, from the time when the colony is being founded, the queens content themselves with reproducing, and leave all remaining tasks to the workers. The workers' area of responsibility embraces the entirety of caring for the brood, building nests, obtaining food and defence.

In view of this abundance of tasks, one may ask oneself whether all workers can carry out all tasks, or whether individual workers are designated specific tasks on the basis of physical or behavioural/ physiological features. This preference of the workers for specific activities has been termed polyethism by Weir (1958a,b). Before we go into the question of polyethism ants in more detail, however, we should firstly describe the characteristics of the various tasks more exactly.

6.2.1   The various activities of the workers

The tasks of ant workers can be roughly divided into two categories of work, i.e. those inside and those outside the nest. According to Otto (1958), in *Formica polyctena,* work inside the nest can be chiefly subdivided in the following manner: 1. the care of eggs and larvae; 2. the care of pupal cocoons; 3. the care of the queens; 4. the care of workers; 5. the disposal of booty in the nest; 6. moving around restlessly without other activity; and 7. remaining in the tunnel system and at the exit of the nest (sentries?).

Firstly, the cleaning of the offspring and licking them with the tongue comes under the care of the eggs and larvae. Another service performed by the workers is the transportation of the eggs and larvae to the part of the nest where they will best be able to develop. This transportation is made easier by the fact that both eggs and larvae adhere to each other, forming clumps. This adhesion is achieved, in the case of the eggs, by means of an adhesive layer on the skin of the egg and, in the case of the larvae, by means of hook-shaped bristles, with which they are covered. Finally, the feeding of the larvae also comes under the care of the eggs and larvae. Among wood ants, this occurs by means of liquid food, which is regurgitated by the workers from their crop, and held in front of the larva's mouth opening, or alternatively by placing the larvae on to insect flesh which has previously been licked and salivated by the workers.

The pupal cocoons are also cleaned and rearranged by the wood ant workers, and it is their job apart from this to help the larvae during the spinning of the cocoons, and to help the young animals when they are leaving the cocoons. This help is necessary for many ant species whose larvae spin themselves into a cocoon to pupate. They are covered with particles of earth whilst spinning, on which they are able to attach the threads, and, after pupating, are freed from the cocoon by the workers. The new young ants, whose cuticula has not yet hardened, cannot usually manage to gnaw their way through the cocoon by their own efforts, and do not possess any chemical means, as do many other insects, of dissolving the cocoon. They therefore have to be released from the cocoon from the outside. This is not the case, however, for all ant species. Young Ponerines of the *Amblyopone* species are able to free themselves from the cocoon independently, gnawing their way through it (Wheeler, 1933). The spinning thread of the wood ant consists of proteins (Brander, 1959; Schmidt and Gürsch, 1971), which are produced in the form of a triple thread out of the labial gland (Schmidt and Gürsch, 1970).

Apart from licking and feeding, care of the queens includes assisting with the storing of the eggs. During the laying of eggs, the wood ant queen stretches her body far up from the ground, whilst the tip of the rear section of her body is touched by the feelers of a worker. The emerging egg is then taken by the worker, licked and then taken to the place where the eggs are being stored. According to the observations of Otto, there is no fixed group of workers which is devoted exclusively to the attendance and care of the queens. There is thus no special 'court', as has sometimes been supposed.

The workers themselves are also cleaned and fed by their own kind, both frequently and abundantly. Ants clean the greater part of their bodies themselves, by drawing the feelers through a cleaning comb on the forelegs and working on other parts of the body with their mouth parts. Reciprocal cleaning among ants takes place by use of the tongue or the mandibles.

Of particular importance is the handing-on of food among the members of a colony. During this, the food is evidently not divided equally among all the members of the colony, as the experiments of Goetsch (1940) already showed: he added to a group of *Lasius flavus* workers, two individuals of the same colony, one of which had been fed with sugar solution treated with blue dye, the other of which had been fed with colourless, but poisoned food. The cuticula of these pale coloured ants is so transparent that it is possible to recognize the coloured contents of the crop in the living animal. The two workers which had been added to the group imparted what they had drunk to the other animals, and it was possible to trace which of the two donors had given them food, either from the blue colour of the contents of the crop, or from the death of the workers. Goetsch discovered during this experiment that the poisoned animals never had blue-dyed contents of the crop. The workers, therefore, received their food from either one or other of the two donors, never, however, from both.

It was possible to investigate the division of the feeding substantially more precisely by means of food marked by radioactivity (Eisner and Wilson, 1958; Gösswald and Kloft, 1956, 1958, 1960a,b, 1963; Kneitz, 1963; Lange, 1967; Schneider, 1966, 1972; Stumper, 1961). It was

found with *Formica fusca* and *Formica pallidefulva* that food
introduced by a single worker was distributed throughout a group of
100 workers after approximately 30 hours (Wilson and Eisner, 1957).
The sugar solution was also distributed with similar rapidity among
other Formicines, the time of year, temperature, size of the group,
saturation level and food supply playing an important part (Gösswald
and Kloft, 1960a,b; Kneitz, 1963). Among the only European species
with replete ants, *Proformica nasuta,* the flow of food is evidently also
regulated according to temperature. At 20°C, the food tends to go to
the storing animals (the replete ants), but at 30°C the food tends to
flow in the opposite direction, from the replete ants back to the rest of
the workers. A similar flow of food was also found among this species
under natural conditions: at cooler times of the year, and when there is
an abundant supply of food, the replete ants tend to fill up, whilst
during hot, dry periods the dominant pattern is the return feeding of
the workers by the replete ants (Stumper, 1961).

According to laboratory experiments by Lange (1967) with the
wood ant *Formica polyctena,* the number of ants receiving food is
determined essentially by the nature of the food source. Food rich in
carbohydrates, i.e. food obtained essentially from visiting aphids, tends
to go to those workers who serve the colony outside the nest, whilst
high-value food, rich in protein, tends to be given to the animals serving
inside the nest, thus benefitting the queens and the brood. However,
protein-rich and carbohydrate-rich foods are, to a large extent, mixed in
the crops of the wood ants, as emerges from the investigations of
Horstmann (1974a), so that the significance of the differential
handing-on of food by the workers serving outside the nest, as
discovered by Lange, remains unclarified.

The research conducted by Schneider (1972) also contradicts the idea
of an equal distribution of food brought into a colony to all the
members. According to this research, there are, in smaller groups of
wood ants (*Formica polyctena*), even in those consisting only of
outside workers, frequently individual ants who only take food and
never hand it on, even when presented with begging behaviour. Other
workers of the same groups, who have themselves received far less food,
are willing to feed others. The result is, therefore, that of the food
brought into the colony, individual workers may receive over 500%
more than others. The varying body sizes of the animals concerned
evidently has nothing to do with this differential intake.

By way of another activity inside the nest, Otto distinguishes the utilization of captured booty among *Formica polyctena*. This activity involves the treating of this booty, brought into the nest by the outside workers, in such a way that it will serve the colony as food. The prey is divided into small pieces, and the edible parts distributed in the colony. Among wood ants, this also occurs occasionally with dead nest mates.

It may sound strange at first to enumerate 'moving around restlessly and without activity' among the various forms of work. In fact, over a third of all workers in a wood ant colony remain idle in the normal way of things (Otto, 1958). This inactivity ranges from short periods of rest to what can be described as the dominant behavioural characteristics of these workers. They seem to have specialized in 'doing nothing'. Otto has observed that the gaster of these workers which tend not to work is particularly distended, which allows one to conclude either that they have a very full crop, or a large amount of fat. It is thus possible that at least a proportion of the idle workers act as living reservoirs for the colony's food economy.

The transition from service inside the nest to service outside does not occur all at once among wood ants, but only after a transitional phase of shorter or longer duration. During this time, the animals gradually move further and further away from the cells in which the queens and the brood are to be found, moving nearer and nearer to the nest exit. It seems as if the young workers need to overcome a reluctance to leave the nest (Dobrzanska, 1959). The first excursions of the young workers out of the nest are thus only of short duration. They return to the nest very quickly and tend to remain at the entrances to the nest, where they greet other returning outside workers with outstretched feelers. It is possible that these workers in the transition period between being inside and outside workers function as sentries, which would also correspond to circumstances among honey bees.

Connected with the transition of wood ant workers from inside to outside service are striking differences in the level of development of the ovaries, as has been indicated by Weyer (1927, 1928) and investigated in more detail by Otto (1958). The ovaries of wood ant workers are similar in their essential features to those of the queens. They consist of a left- and a right-hand group of ovarioles, which open into short oviducts. The left- and the right-hand oviducts open into a common excretory duct leading to the outside of the body and situated at the tip of the rear section of the body (Fig. 6.2). The eggs mature in the ovarioles, surrounded by nutrient cells, and become joined to the oviducts as they mature. Whereas between 66 and 83 ovarioles in each

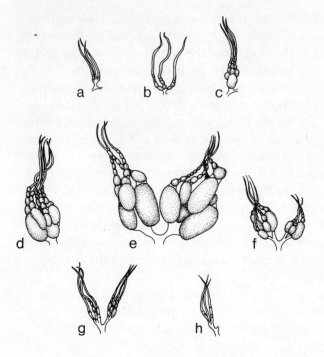

Fig. 6.2   The various stages of development and degeneration of the ovaries of
wood ant workers. (a) ovariole of a freshly hatched worker; (b-e) the stages of
development up to that of the full development of the eggs; (f-h) stages of
degeneration (after Otto, 1958).

ovariole group could be counted among queens, there were only 2-5
ovarioles on either side among the workers (Otto, 1958).

Among freshly hatched workers, the ovarioles are still very thin and
undeveloped. The ovarioles develop with the increasing age of the
workers; it finally becomes possible to distinguish the nutrient chambers
of eggs, which, particularly in the oviduct area, grow extraordinarily
large (Fig. 6.2). Some of these eggs even outdo in size those of the
young queens, but are not laid as a rule. They are reabsorbed by the
body and shrivel — depending on the time of year — together with the
rest of the eggs (Fig. 6.2).

If one compares the condition of the ovaries of inside and outside
workers, one can see that the ovaries of the outside workers are
substantially more degenerated than those of the inside workers (Weyer,
1927, 1928; Otto, 1958; Hohorst, 1972). It is possible that ovarian

degeneration affects the transition of the workers from inside to outside work; however, it has not been possible to prove this up to now. Nevertheless, apart from the fact that the ovaries are degenerated among those wood ant workers which clean the nest of waste and those concerned with the building of the nest, Otto concludes that these ants belong to the outside workers.

The typical outside tasks of the wood ants are aphid visits, bringing captured prey into the nest, the transportation of nest mates and bringing in nest materials. According to the observations of a number of authors, a further division of labour is possible in connection with aphid visits. Adlerz (1886) made this observation with wood ants; 'aphid milkers' handed on the food they obtained to other, usually larger workers, who for their part took the food into the nest. On the basis of this, he concluded that there was a division of labour between 'milkers' and 'transporters'. King and Walters (1950) made similar observations among the North American wood ant *Formica melanotica* Emery (=*Formica obscuripes* Forel). Eidmann (1927, 1929a) found a special group of workers among *Lasius niger,* which, in the spring, would spend the whole day with the aphids, but were not observed to milk them. In the evening they returned to the nest and ran to the same group of aphids again the next morning. Eidmann, on eight days of observation, experimentally noted the ability of these animals to return to the same place by marking them with coloured spots. From the observation of these aphid visiting ants, who do not themselves milk aphids, he concluded that they served a special sentry function. In summer, when the aphid swarms are larger and the nights become warmer, the *niger* workers transport aphid honeydew into the nest, mainly after dark. The ants avoid bright daylight at this time, but in spite of this, Eidmann was able to observe sentries among the aphid swarms even in bright sunlight. Herzig (1938) was unable to confirm this in his investigations, however. Sentries have also been reported among wood ants (Elton, 1932; Hölldobler, 1944; Wellenstein, 1957), but Otto was unable to detect these in his observations. Otto was also unable to confirm the existence of the transporting group of animals for honeydew in his marking experiments. All things considered, therefore, clear and unequivocal evidence is still lacking to justify the subdivision of aphid visiting activity into further sub-tasks.

A similar situation applies to the insect hunt: Otto discovered that wood ants hunting for prey were smaller than the other workers, who transported the captured prey into the nest. On the basis of this, he concluded the existence of a possible division of labour into hunters and transporters of captured prey. Büttner (1974a,b) and Horstmann (1973a) on the other hand, discovered that the fighters and transporters were identical as a rule.

After this overview of the various activities of workers, the question now arises as to the factors which predispose the workers for their respective tasks. We know in the case of honey bees, that age, and probably also the increasing influence of the juvenile hormone from the *corpora allata* of the young workers is responsible for the activity of the animals (Rutz *et al.*, 1976). The young bees take on all tasks appropriate for workers in their turn, as they coincide with their continuing development. They begin by cleaning the empty cells, and then become nurse bees when their royal jelly glands have developed, and feed the young larvae. The royal jelly glands then recede, whilst the wax glands become active, turning the workers into builder bees, who see to the construction of the honeycomb. After the wax glands have also reformed, the bee workers then become sentries, serving at the entrance to the colony. From here they later undertake their first orientation flights in the vicinity of the colony, and these become increasingly more extended, until finally they bring in pollen and nectar as flying bees.

Amongst ants, several factors affect the activity of the workers. We shall deal with these factors separately, firstly the influence of age, then the influence of polymorphism, i.e. the influence of differential physical characteristics on polyethism.

6.2.2 Age polyethism

As far as the influence of age on polyethism is concerned, we know that amongst ants, freshly hatched workers generally serve inside the nest first, becoming occupied outside the nest only later. The changeover from service inside to service outside occurs at widely varying times, not only among different ant species, but also among different workers of the same colony. Age determined polyethism has so far been investigated among *Manica rubida* (Ehrhardt, 1931), *Myrmica scabrinodis* (Weir, 1958a,b), *Messor* (Goetsch, 1930), *Formica polyctena* (Otto, 1958) and *Formica sanguinea* (Dobrzanska, 1959).

Among *Formica polyctena,* the transition from inside to outside service does not, as a rule, occur before the 40th day after hatching. Workers that hatched at the beginning of the brood period (beginning of June) may thus come to serve outside the nest in the same year.

According to Otto (1958), there are, however, also animals which become outside workers much earlier, or much later; some workers even remain working inside the nest all their lives. This shows that the transition from inside to outside service, and therefore also development and degeneration of the ovaries of wood ant workers does not occur at any fixed time. It depends on external factors, such as diet and on the presence of outside workers. If in other words, one isolates freshly hatched workers from the rest of the colony, then the ovaries remain undeveloped; the same effect can be observed among workers which have not received a protein diet. Among other *Formica* species, such as *Formica rufibarbis,* however, it was not possible to confirm the effect of old ants and a flesh diet on the development of the ovaries (Hohorst, 1972).

Workers may participate in social tasks very early in their service inside the nest. On the third day they are able to clean their companions and pupae, after four days they are in a position to feed and care for eggs, larvae, pupae and queens. This still does not mean, however, that the workers do so. At what age these activities are actually carried out, and indeed if they are at all, varies greatly from one individual to another. It is also not possible with wood ants, as it is with bees, to correlate a particular activity inside the nest with the level of development of specific glands.

Dobrzanska (1959) came to similar conclusions regarding the slave-making predator ant *Formica sanguinea.* In general young workers go to serve outside the nest as early as a few days to three weeks after hatching; however some leave the nest as early as one or two days after hatching, while still others work inside the nest all their lives.

Among the Myrmicines also, Ehrhardt (1931) and Goetsch (1930) found that the freshly hatched workers initially remain in the nest. In the case of *Myrmica ruginodis,* Weir (1958a,b) distinguishes various groups of workers according to the behaviour of the animals, these groups probably representing various age levels. He distinguishes between 'nurses', who sit on the brood, 'domestics', who tend to remain in the vicinity of the brood and the food searchers working outside. The nurses withstood longer periods without food, laid the most eggs, had the most difficulty killing muscid larvae and were the worst at finding buried prey. In these respects, the outside workers stood in contrast.

Cammaerts-Tricot investigated *Myrmica rubra* workers by subdividing them according to their degree of pigmentation, from pale, freshly hatched ants to fully pigmented animals. She was able to show that the quantity of pheromone present in the mandible gland increases with the age of the animals. This held to some extent also for the reservoirs of the Dufour glands, which were only completely full in workers with a pigmentation level of three. This age-dependent, variable supply of pheromones went hand in hand with an increased reaction on the part of the workers to these signalling substances (Cammaerts-Tricot, 1974c). The trail following reaction increases in relation to the age of the workers, as does the quantity of trail substance in their poison glands (Cammaerts-Tricot and Verhaeghe, 1974).

Wehner *et al.* (1972) found a close connection between the level of development of workers and their favoured work in the colony among the desert ant *Cataglyphis bicolor.* Among the workers of this species, not only the ovaries, but also other internal organs, such as the labial glands and the fatty compounds, were developed to varying degrees. According to the condition of these organs, Wehner was able to draw up a development sequence, in the course of which not only would these organs progressively degenerate, but the workers would also change their functions within the colony. These organs are most reduced among the inside workers, more developed in those animals which tend to be carried ('carried stage') and those which occasionally serve as storing animals, still more developed among the builder workers ('digger stage'), the hunting animals ('hunter stage') and finally, most developed among the carrier ants ('carrier stage'). The sequence of these stages seems to be fixed, but not, however, the speed with which the ants progress through them.

6.2.3   Polymorphism and polyethism

The workers of a colony can differ from one another considerably in relation to body size and other morphological features (polymorphism). One can see this, for example, in the case of leaf-cutter ants, where the largest animals of a colony can be many times greater in size than the smallest (*see* Fig. 6.6). Other morphological variations concern peculiarities of form in the head, mandibles or gaster. In cases where differences in the external structure of workers are linked with behavioural variations concerning the kind of work favoured in the colony, we speak of 'caste polyethism'.

One subcaste already mentioned as existing among a number of ant species is that of the soldier ants, a subcaste usually represented by particularly large workers with over-large heads and powerfully developed mandibles. We can in fact only speak of soldiers in the true sense when these workers are not connected with the rest of the workers by means of any intermediary form. The workers, in accordance with their body size, are allotted places in a continuous sequence, as major, medial, minor and minimal workers.

In accordance with their particular morphological features, the soldiers and major workers serve especially for the defence of the colony. This goes for the major workers of the South American leaf-cutter ants as well as for the soldiers of the Aneuretines and the harvester *Pheidole* species (Fig. 6.3). These ants all differ from other

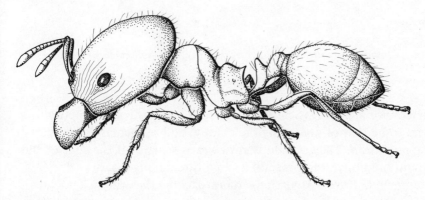

Fig. 6.3   *Pheidole militicida* soldier (after Creighton and Creighton, 1959, adapted).

workers mainly in the fact that they are larger and stronger, possess larger and more powerful mandibles and defend more readily and effectively (Creighton and Creighton, 1959).

In contrast to the type of soldier as represented in the *Pheidole* genus, the soldiers of army ant species and those of the Sahara dwelling species *Cataglyphis bombycina* differ from the other workers not only in their body size, but also in the particular form of their long and unusually pointed mandibles (Fig. 6.4). These mandibles are hardly used for peaceful work, but, for this very reason, are all the more effective as weapons. Délye has also observed in connection with this that *Cataglyphis bombycina* soldiers are in practice only active at times when the colony is being defended. This is also the case for the soldier ants of the South American army ant genus, *Eciton,* who constitute about 2%

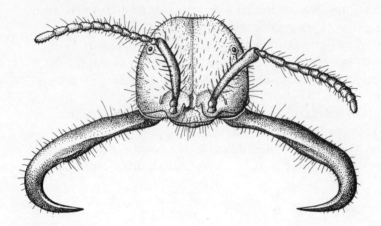

Fig. 6.4   Head of an army ant soldier of the species *Eciton hamatum* (after Topoff, 1972).

of the colony members (Schneirla, 1971). Their particular function appears to be that of defending the colony, and particularly the queen, during migration. They station themselves as sentries at the edge of the army and hold open their mandibles in a threatening manner (Rettenmeyer, 1963). Any stranger approaching the army is attacked by the sentries. The queen of the exclusively monogynous army ants is watched by a particularly large number of soldiers, who when in danger, not only defend the queen vigorously, but will also lie protectively over her (Schneirla, 1971). Topoff (1971) made similar observations among *Neivamyrmex nigrescens,* another South American army ant species.

The soldiers of *Colobopsis, Hypercolobopsis, Pseudocolobopsis,* and *Myrmaphaenus* serve to defend their colonies in a completely different manner. Forel was the first to describe the function of *Colobopsis truncatus* soldiers in 1874. This species, which occurs in Central Europe, nests in the branches of living trees. Connection with the outside world is made only through a single hole, which is so small that the inhabitants are only just able to pass through it. The soldiers which occur in small numbers in a colony, defend it by blocking the entrance (Fig. 6.5). They have a broadened head shape, which is flattened off in front and fits exactly into the entrance hole. Inasmuch as the head is visible from the outside, it is so structured and coloured that it is barely distinguishable from the surrounding bark. If the entrance hole has become larger than the head of a soldier, it can be made smaller by means of a box-like mass. It also occurs sometimes, however, that a larger entrance hole is simultaneously blocked by a number of soldiers. The soldiers remain for a long time at their posts in the entrance to the nest and only open the entrance when a worker, probably recognized by its scent, knocks it on the head with its feelers.

The entrance to the nest is closed in the same manner by soldiers

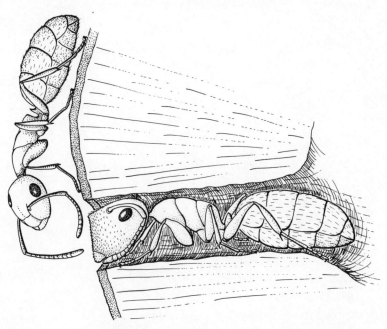

Fig. 6.5   Colobopsis truncatus soldier, blocking the entrance to the nest with its head, whilst a worker arrives at it (after Wilson, 1971 — from Szabo-Patay, 1928, adapted).

among the *Cryptocerus texanus.* In order to block the nest entrance, the soldiers use not only their specially formed head shields, but also a portion of the thorax. A worker returning to the nest makes the soldier 'keeping watch' lie flat on the surface of the nest entrance, allowing sufficient room between its back and the roof of the nest tunnel as an entrance (Creighton and Gregg, 1954; Creighton, 1963).

The modifications of these soldiers as an adaptation to their particular sentry function have been described by Wheeler (1927) as phragmosis. Wheeler (1901, 1927) also describes queens with phragmotic heads among the species *Crematogaster cylindriceps* and *Colobostruma leae. Colobopsis* queens also have phragmotic heads. By way of further variation, Brown discovered in 1967 phragmotic gasters in the queens of *Pheidole embolopyx,* a species which inhabits the Amazon area, but we have no information as to what function the phragmotic rear section of the queen's body served at which of the colony's phases of development.

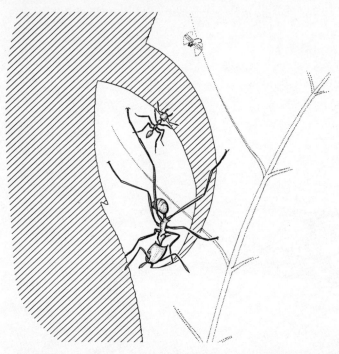

Fig. 6.6   Small worker of the leaf-cutter ant species *Atta cephalotes,* accompanying a medium-sized nest mate bringing in a leaf, and keeping off parasitic flies (Phorides) (after I. and E. Eibl-Eibesfeldt, 1967, adapted).

Eibl-Eibesfeldt (1967) was able to observe one of the preferred activities of particularly small workers among the leaf-cutter ant *Atta cephalotes.* These remarkable minute animals, described as minima workers, are not only the most effective fungus tenders in the colony; they also protect the larger workers from the attack of parasitic flies of the Phoridae group. The defence by the tiny ants is particularly important when the larger workers are cutting leaves and are themselves defenceless. The minima workers station themselves with opened mandibles by their leaf-cutter nest associates and snap at the approaching flies (Fig. 6.6). They then ride back into the nest as sentries on top of the piece of leaf which has been removed.

We have already acquainted ourselves with a particularly clear form of polyethism as it exists between the workers and replete ants of the honey ants. Here, the replete ants serve exclusively as living food reservoirs and do not take part in the other activities of the nest. It seems to be a common feature of all ant species with replete ants that only the largest workers become storer animals (Wilson, 1963).

Among some ant species, it is possible to recognize inside and outside workers by their different body sizes. Among the weaver ant *Oecophylla longinoda,* and among *Daceton armigerum*, the smaller workers remain in the nest, whilst the larger ones procure food (Ledoux, 1949b; Weber, 1949; Wilson, 1962a). The circumstances are exactly the reverse among *Formica obscuripes.* Here, it is the smaller workers who procure food, whilst the larger ones remain in the nest (King and Walters, 1950). According to the laboratory observations of Dobrzanska on *Formica sanguinea* (1959), here also it is the larger workers who occupy themselves with the care of the brood in the nest. Otto (1958) and Kneitz (1967) undertook extensive measurements with the wood ant *Formica polyctena*, and were unable to establish any significant difference in size between inside and outside workers.

When comparing the body size of outside wood ant workers, who pursue a variety of tasks, Adlerz (1886) found that smaller workers 'milked' aphids, whilst larger workers brought the food into the nest. According to Alpator and Palenitschko (1925), the workers of *Formica pratensis* discovered in an aphid colony close to their own nest, were smaller than the average size of those within the nest. Otto (1958) was also able to observe a difference in size in *Formica polyctena* between aphid visitors, collectors and hunters. This was also confirmed by Horstmann (1973b), who observed that larger *Formica polyctena* workers tend on average to go further away from the nest, to hunt and gather food on the ground and to move loads, more often than smaller workers which, according to Horstmann, keep closer to the nest, climb trees more often and correspondingly bring in more honeydew.

6.3   Causes of caste differentiation

The origins of caste and sex differentiation vary substantially among different ant species. As has already been stated in the introduction, among ants, and among Hymenoptera in general, females result from fertilized, diploid eggs, whilst males originate from unfertilized, haploid eggs. A queen able to produce males as well as females must thus be capable of laying fertilized as well as unfertilized eggs.

It has been found in a range of ant species, however, that workers are also able to lay eggs which will develop, particularly when the colony is without a queen. Since the workers are not inseminated – this is not contradicted by the few examples of inseminated 'workers', such as are to be found among *Hypoponera eduardi* (Le Masne, 1956a), *Technomyrmex albipes* (Weyer, 1936) and among certain *Rhytidoponera* species (Whelden, 1957, 1960; Haskins and Whelden, 1965), if one defines 'queens' and 'workers' functionally – we must assume that only males result from the eggs of these workers. This is in fact true in the overwhelming majority of cases; interestingly, however, there are also exceptions.

As early as the last century, we find observations according to which, not only males, but also workers and queens resulted from the eggs of leaf-cutter workers (Tanner, after Ledoux, 1954). Reichenbach (1902) and Crawley (1912) found that among *Lasius niger,* workers could develop from workers' eggs. Further examples of this were found by Ledoux (1949a,b) among workers of the weaver ant *Oecophylla longinoda,* by Otto (1958) among the wood ant *Formica polyctena,* by Soulié (1960a) among workers of a *Crematogaster* species and by Cagniant (1973) among *Cataglyphis cursor.* In 1945 Haskins and Enzmann showed that workers could also develop from the eggs of non-inseminated queens of species of *Aphaenogaster.*

In 1960a,b Soulié examined the various types of eggs laid by females as well as workers in *Crematogaster* colonies, and was able to distinguish three types of egg: unfertilized eggs with haploid chromosome composition, unfertilized eggs with diploid chromosome composition and finally fertilized diploid eggs. Unfertilized haploid eggs are laid by queens as well as workers and develop into males. Unfertilized diploid eggs are only laid by workers and result in full females, whereas only workers develop from the fertilized diploid eggs of the queens.

This shows that ant workers are in fact able to lay diploid eggs. How this is possible without insemination was investigated by Ledoux (1949a,b) with *Oecophylla longinoda.* In this species, the queens lay eggs which hardly vary at all in size. These eggs may be inseminated and then develop into females; they may also remain unfertilized, however, and result in males. *Oecophylla* workers, on the other hand, lay two different types of eggs, varying in size. The larger eggs, with a diameter of 1 mm, are laid before their maturation division, and are diploid as a result. The first embryonal cell division of these eggs, however, is a reduction division, making haploid out of diploid eggs. Only males develop from these eggs, therefore. The smaller eggs have a diameter of 0.6 mm; they are held back in the oviduct of the worker until the first maturation division (the equation division) is completed. As a result of this they are diploid, and remain so after the subsequent divisions. From these eggs develop full females exclusively, as from the diploid eggs of *Crematogaster* workers.

Diploid eggs may, therefore, result in workers as well as full females. The reasons why workers develop from some diploid eggs and full females from others are evidently not of a genetic nature. Three other factors were found among ants in support of this, having an effect on caste differentiation, these being the presence of the queen, trophogenic and blastogenic factors. The trophogenic factors embrace all those concerned with the feeding of the larvae, whilst the blastogenic factors concern the particular features of the egg plasma. As regards the various subfamilies of ants, knowledge of the origins of caste differentiation is at different levels. We still know almost nothing of the origins of caste differentiation among Myrmeciines, Ponerines (with the exception of *Odontomachus haematodes*) and Dolichoderines, but we do know something, however, of caste differentiation and regulation among Formicines and Myrmicines. This has been dealt with essentially by two working groups. Bier and Gösswald in Germany investigated mainly the *Formica* genus, whilst Brian *et al.* investigated the Myrmicine genus *Myrmica.*

Among many species, the presence of the queens has the effect of inhibiting the number of full females produced. If, in other words, one removes the queen after the eggs have been laid, in species of *Odontomachus* (Colombel, 1971a, 1972a), *Monomorium* (Peacock *et al.* 1955), *Myrmica* (Brian and Carr, 1960) and *Formica* (Gösswald and Bier, 1954b; Bier, 1956), more full females develop from the larvae than would have done if the queen had been present. In the case of *Monomorium pharaonis,* Peacock *et al.* (1955) were able to show that colonies without a queen rear sexual animals from existing larvae, that is, both females and males. If queens, or even only non-inseminated females are present (Petersen and Buschinger, 1971a), then the formation of the sexual animals does not occur.

In *Formica polyctena,* which habitually harbours a large number of queens in its colonies, 'physiological absence of queens' is, according to the results of Gösswald and Bier (1954b), the prerequisite for the production of sexual animals. In the early spring, after awakening from the winter rest period, the queens gather together close to the workers on the surface of the nest and warm themselves in the sun. This 'sunning period' is for a limited time: afterwards, the queens lay their first eggs in the top of the nest and later wander back into the lower chambers which are built into the cold mineral-deposited ground. After the first larvae have hatched, only a very few queens are still to be found in the top of the nest; the queens are thus spatially separated from the brood. This separation of the queens from the brood is that described by Gösswald and Bier (1954b) as the 'physiological absence of queens'. It is a necessary prerequisite for the formation of the sexual animals, which develop from the first eggs of the year laid by the queens. If one prevents the queens from moving away from the brood, then the proximity of the queens leads to only workers being produced from these first eggs of the year.

A similar effect has also been found among the species *Formica nigricans,* which only keeps one or a few queens in its colonies. Here also, the presence of the queen inhibits the development of the larvae into full females. This effect of the presence of the *nigricans* queens can, however, be compensated for by a correspondingly large number of workers. If the ratio of workers to queens is more than 600:1, the same number of sexual animals will develop from the larvae whether the queens are present or not (Gösswald and Bier, 1953b).

The presence of the queens in the colony frequently not only hinders the development of the larvae into full females, but also suppresses the egg-laying of the workers. This is not the case, however, for those species where the caste differences are so marked that the workers are completely incapable of laying eggs, such as in the *Pheidole* and *Solenopsis* genera, among *Tetramorium caespitum,* in the army ant genus *Eciton* and among *Monomorium pharaonis.* The caste differences are not so marked among the *Lasius* and *Camponotus* genera, where only the smaller workers remain sterile, whilst the particularly large workers are still able to lay eggs (Bier, 1958).

The inhibition of the fertility of workers by the presence of the queens has been demonstrated for the *Leptothorax* (Bier, 1954a), *Myrmica* (Mamsch and Bier, 1966), *Plagiolepis* (Passera, 1965) and *Formica* (Bier, 1956) genera, by the fact that far more eggs are laid in colonies without queens than those with queens. With a nest temperature, for example, of 20.5°C, 50 *Leptothorax unifasciatus* workers had laid 140 eggs after 42 days; when the queens were present, however, only 60 eggs (Bier, 1954a) were laid. In naturally formed colonies of this species, it is probably the queens alone who lay the eggs, not the workers (Bier, 1956).

According to Wheeler (1923), however, the egg-laying of the workers, at least in monogynous colonies, is not entirely suppressed; it plays on the contrary an important part in 'queen-regulated' colonies, particularly in the production of males. This has also been confirmed by Ehrhardt (1962) for the polygynous species, *Formica polyctena*. He discovered that workers of this species also lay eggs when the queen is present; these are larger than the eggs of the queens and can be distinguished from them as a result. Only males develop from these worker eggs. Apart from this phenomenon, there are, according to the investigations of Ehrhardt (1970), in *Formica polyctena* colonies a number of queens whose sperm supply has run out, or some which, although they shed their wings, have not been inseminated (*see also* Gösswald and Schmidt, 1960), and which have therefore been taken in by the colony. These queens thus lay only unfertilized eggs, to which, quite probably, the greater proportion of the males of the colony are developed. Weyer (1928) on the other hand, maintains that the reproduction of workers in colonies with queens is practically insignificant.

It is known that in a number of species the presence, not only of the queens, but also the larvae can suppress the fertility of the workers. Mamsch (1965, 1967) came to this conclusion as a result of experiments with *Myrmica ruginodis*. By adding larvae to queenless colonies, he was able to prevent the workers from laying eggs, even in those cases where the workers had already begun to lay the eggs.

The influence of the queens on caste differentiation among the brood and their effect on the fertility of workers is also very clearly noticeable among honey bees. Here, it has been known for over ten years that the effect of the queens manifests itself by chemical means, that is by the secretion of their mandible glands. This secretion was previously known by the collective term 'queen substance' (*see* Butler and Simpson, 1958). Nowadays this term is usually applied to only one of its main constituents, trans-9-oxodecane acid-2. Another main constituent of this secretion is, among at least 30 other substances, trans-9-hydroxydecane acid (Callow *et al.,* 1964). The secretion of the mandible gland is distributed over the food in the colony. The trans-9-oxodecane acid has the effect on the workers of inhibiting the reproduction of queens. It, together with the trans-9-hydroxydecane acid, inhibits the ovarian development of the workers.

We still do not know exactly in what way ant queens exert their influence on the caste differentiation of the brood and the fertility of the workers. For some time, there have been two hypotheses about this. According to the first hypothesis, ant queens exercise their influence in the manner of the honey bees, by means of pheromones (Karlson and Butenandt, 1959; Brian, 1965a,b); according to the second hypothesis, represented essentially by Bier, the effect of the ant queens can also be explained without recourse to pheromones.

In support of the first hypothesis are the investigations of Carr (1962), which show that, in *Myrmica,* even the presence of dead queens, if they are replaced regularly, inhibits the growth of larvae and thus reduces the number of full females produced. Substances extracted from queens, however, showed no effect, regardless of the form in which they were presented to the workers. Carr concluded from this that the substance sought was volatile, and only present in small quantities.

Brian and Hibble found in 1963 that the secretion from the mandible glands of *Myrmica* queens inhibited the growth of the larvae. Even the extracts from the whole heads of queens were effective in ethanol, albeit only half as strongly as living queens. Then in 1969, Brian and Blum analysed the extracts from the heads of fertile *Myrmica rubra* queens into various classes of substances, and tested their effects on the growth of the larvae. They discovered that only one of the fatty acids found in the extracts had an effect.

The investigations of Colombel (1971a, 1972a) also support the pheromone hypothesis. He found that among the Ponerine species *Odontomachus haematodes,* the formation of workers was accelerated and the production of full females suppressed by contact with the queens, or even their scent, by the workers. The removal of the queen resulted in the workers being prevented from eating eggs, as a result of which more workers, but also full females, were produced.

According to Bier (1954a, 1956, 1958) and Mamsch (1965, 1967), the production of inhibiting pheromones by ant queens should certainly not be ruled out, but they do not regard it as necessary to put this forward as the explanation of the inhibiting effect which the queen has on the fertility of the workers and the reproduction of sexual animals. Instead of this, these authors adopt the hypothesis that the workers produce a fertility enhancing, 'profertile' substance in their feeding glands, and that this substance tends to be fed to the queens. If queens are absent in the colony, then the profertile substance is handed on, with more concentrated effect, to the larvae, who then develop in greater numbers into sexual animals. As long as these two main receivers of the profertile substance are present in the colony, the substance produced by the workers will be taken by them, and the workers themselves will remain infertile. If, on the other hand, queens as well as larvae are both absent in the colony, then the larger workers are next in line for the profertile substance, becoming fertile and laying eggs as a result.

The observations made with *Myrmica* (Mamsch, 1965, 1967) and also with *Leptothorax* (Bier, 1954a) support this hypothesis; according to these, larvae also exert an inhibiting effect on the egg-laying of the workers. The larvae are, according to this hypothesis, the preferred receivers of the profertile substance before the workers, and their receipt of it prevents the workers from becoming fertile. We know furthermore from feeding experiments with radioactive marked substances (Gösswald and Kloft, 1960a,b) that in the *Formica* genus, where there is a sufficient number of workers caring for the queens (at least 30), the queens are fed with pure glandular secretion — this has also been demonstrated in the case of *Monomorium pharaonis* (Buschinger and Kloft, 1973) — whilst the workers exchange the content of their crops among themselves. The observations of Bier with *Formica nigricans,* where the influence of the queens on the development of the larvae abates if a sufficient number of workers is present, may also be interpreted in light of this hypothesis. In this case, there would be a surplus of profertile substance present due to the large number of workers, this surplus would then be fed to the second main recipients of the substance, the larvae, whereupon these would develop in greater numbers into queens.

The composition of the secretions of the various glands involved in brood feeding and as purveyors of the profertile substance respectively has been investigated by Paulsen (1966, 1969, 1971). In connection with this, he tested the labial, propharynx and postpharynx glands of *Formica polyctena,* and found in the postpharynx gland a specific enzyme which broke down sugar. This supports the hypothesis that the postpharynx gland is mainly concerned with the digestive function. The yellow-coloured secretion of the postpharynx gland, rich in lipids, is, as Paulsen was able to show, transferred to other workers, the queens and the larvae. The fatty substances of the postpharynx gland come very probably from the fat compounds. It is possible, moreover, that those secretions — which are important for breeding the sexual animals, and possibly also the caste determining (*see below*) – and profertile secretions, are produced in the postpharynx gland. Finally, the labial glands contain mainly feeding secretions, mostly carbohydrates originating from reabsorbed food and mobilized reserve substances.

The queens of socially parasitic ants, which live in the colonies of other species, also seem to display a fertility inhibiting effect. Queens such as those of the workerless parasitic species *Myrmecia inquilina* inhibit the ability of their host queens to lay eggs, although they do not behave towards them in a hostile manner (Haskins and Haskins, 1964). The relationship between the parasitic ant *Plagiolepsis xene* and its host species *Plagiolepsis pygmaea* is reciprocal. As in the mixed *Myrmecia* colonies, here too the queens of both species live together without any observable hostility. Whereas the presence of the *xene* queen inhibits the fertility of the *pygmaea* queen, the presence of the *pygmaea* queen, however, also inhibits the ability of the *xene* queen to lay eggs (Passera, 1966).

In the case of *Plagiolepsis pygmaea,* the fertility inhibiting effect of the queen certainly does not lie either in the emission of profertile substances by workers, or in the effect of queen pheromones (Passera, 1969). Extracts of queens and dead queens showed neither an attraction nor any other effect on the workers. After further scent and visual stimuli had been ruled out, the inhibiting effect was clearly traced back to the tactile stimuli of the inseminated, living and moving queens without wings. The inhibiting effect could be exercised only by wingless females in this case, not females with wings.

Apart from the presence or absence of the queens among wood ants, trophogenic factors also affect caste differentiation among the larvae, that is, factors connected with the nutrition of the larvae. Workers are responsible for the feeding of the larvae, doing so not only from the contents of their crop, but also with the secretions from their royal jelly glands, labial and postpharynx glands. These glands of the workers are directly able, after the winter rest period, to expedite the development of the larvae from fertilized eggs into queens. This emerges from experiments with *Formica pratensis.* If one gives workers of this species eggs to care for after the winter rest period, then full females develop from these eggs if they are fertilized. If, however, they are given similar eggs after they have been kept for about eight weeks at 24°C, then they rear workers from these eggs (Gösswald and Bier, 1953b).

In addition to the physiological condition of the ants feeding the larvae, their number, and thus also the quantity of their glandular substances, are also significant in their relation to the caste differentiation of the brood. Below a specific minimum number of workers caring for the brood, irrespective of their physiological condition, the effect among *Formica pratensis* is that only workers result. This is only the case, however, until the final caste determination of the developing larvae. If eggs are given to a group of less than 25 workers able to care for them directly after the winter rest period, and if these larvae at 72 h old are then given to the care of a larger group of workers in identical physiological circumstances, then only workers will develop from these larvae, not full females. The caste of these larvae is finally determined after 72 h, and the conditions which would otherwise lead to the development of full females are no longer effective after this time. The reverse of this experiment does not, however, succeed in producing the reverse effect: if one removes larvae being cared for by larger groups of workers into smaller groups, the larvae do not then develop any further (Gösswald and Bier, 1953a).

By way of a third factor among wood ants, the egg type also affects the caste differentiation of the larvae (Bier, 1954b). One can distinguish summer and winter eggs among the *Formica* species. The winter eggs are laid directly after the winter rest period by the queens. These eggs contain more RNA, and their periplasma — especially at the rear pole of the egg — is enlarged into the form of a pole plasma. The summer eggs, which are laid during the rest of the year, are recognizable by the fact that their 'protein yolks' are larger and more densely packed than in the winter eggs.

The caste of the larvae, however, is still not determined by the varying egg types. With the appropriate care by the workers, full females as well as workers can develop from both types. The winter eggs, however, develop more easily into full females, and summer eggs develop more readily into workers. Gösswald and Bier (1954a) thus speak of a 'predisposition' of the type of egg for a specific caste. When the predisposition of the egg type was set against the physiological condition of the workers caring for them, intermediary forms between full female and worker emerged under experimental conditions which had not hitherto been found under natural conditions. These 'intercastes' formed more readily from the summer than the winter eggs, and all the transitional features emerged in the formation of the thorax and the wing stump.

In the natural environment, all three factors affecting the caste differentiation of wood ant workers work together in the regulation of caste. The winter eggs predisposed to develop into sexual animals develop at a time when the workers were physiologically able to breed sexual animals. Added to this is the behaviour of the queens, who, after the first eggs have been laid, retire into the cooler parts of the nest, resulting in their physiological absence, another condition for the breeding of sexual animals. The result of these three factors working together, is that among our wood ants, exclusively sexual animals are produced from the first brood of the year. During the course of the rest of the year, among the wood ant *Formica polyctena,* only workers develop. Among *Formica nigricans,* on the other hand, at least in the Würzburg area (Bier, 1956), in July or August a second brood is produced, composed of sexual animals mixed with workers. This is clarification that freshly hatched workers as well as those workers directly after the winter rest period are able to rear a brood of sexual animals. One should add that among *Formica nigricans* the different types of egg play a less important part than among *Formica polyctena* (Gösswald and Bier, 1954a).

The investigations of Brian and his colleagues in England were conducted with the species *Myrmica laevinodis* (*Myrmica rubra* L. according to Collingwood, 1958), *Myrmica rubra var. microgyna* and *var. macrogyna* (*Myrmica ruginodis* NYL. according to Collingwood, 1958). The same essential features affecting caste differentiation among wood ants were also shown to exist among these species. Here also, the effect of the queen and apart from this, trophogenic and blastogenic factors were important for the development of a fertilized egg into a worker or full female as the case may be. This only held, however, in the general sense. Variations can be seen between the *Formica* and *Myrmica* genera in certain fundamental details.

In contrast with *Formica,* a direct effect on the caste of the developing larvae by the workers feeding them could not be show to exist, but it was possible to show an indirect effect. The species *Myrmica ruginodis* NYL., like many other Myrmicines, maintains larvae throughout the winter. This passing of the winter time in larval form is an important stage in the development of young queens. A necessary precondition for a larva to pass the winter in this form is its entry into a nest phase; the occurrence of this phase depends on nutrition. We know from the investigations of Weir (1959), that in natural colonies workers only have the ability to effect a rest phase in the development of the larvae in the autumn. During the course of the brood period, the workers have taken in increasing amounts of protein, which affects the nutrition of the larvae. Among larvae in the rest phase, the ratio of nitrogen to carbon in the fat compound is higher than among workers not in the rest phase. Weir (1959) was able to demonstrate the connection between the rest phase of the larvae and the nutritional condition of the workers caring for them by feeding workers with increasing amounts of protein in the spring. As a result of his doing so, the larvae being cared for by these workers were induced prematurely into a rest phase in their development.

The rest phase in larval development results in their passing the winter in larval form. The phase may be counteracted by a rise in temperature and by the larvae being cared for by the 'spring workers'. Both conditions are fulfilled in the natural colony in spring. The cooling down of the resting larvae in the winter leads to them achieving later a greater growth rate, which is important for the development of the larvae into full females, particularly in the critical period of the third larval stage (Brian, 1956a). Future queens ready to pupate have an average weight of 8 mg, whilst the corresponding developmental stage of future workers shows an average weight of only 4.5 mg.

The last larval stage of *Myrmica ruginodis* is the most critical for the determination of caste. This is the decisive moment as to whether the larva will develop into a full female or a worker. Brian (1954, 1955b, 1956a, 1957) thus investigated this state particularly thoroughly. Not being able to address the question of the individual intermediary stages independently of the absolute size of the developed animals, he set them in relation to morphological changes which precede them in the larval stage. First the brain moves from the head to the thorax, then the leg structure segments into two, and later three parts. The caste of the larva is not determined until the developmental stage has been reached where the brain has travelled half the distance to the thorax. The determination of the worker caste can ensue, however, when the brain has travelled 60-80% of its way. The determination of caste for the full female, on the other hand, can only take place when the brain is situated in the thorax and the leg structure has three segments. If the larvae are left hungry in the time between worker and full female caste determination, then intermediary forms between the two are produced.

Although the absolute size of the larvae is a significant factor in their development to full females, or workers, there are nevertheless instances of overlapping among the larvae which pass the winter in the larval stage. According to Brian (1955b), the largest larvae, from which workers develop, weigh 1.1 mg, the smallest, which become full females, weigh 0.7 mg. In spring the growth rate of the future queens is considerably faster, so that the weights of the fully grown animals of the two castes are indistinguishable. Connected with the varying rates of growth are the different developmental conditions of full females and workers among *Myrmica ruginodis*: workers are here produced through the fact that a portion of the structures which develop in the full females are inhibited in their development in future workers. The structures are made to divide into two parts, the ventral and the dorsal. The dorsal structures behind become wings, ocelli and gonads in their later development. The ventral structures, on the other hand, develop into legs, mouth parts and the central nervous system (Brian, 1957). Whilst in the case of full females both groups of structures develop fully, the dorsal structures are inhibited from developing in the workers.

It is already clear from this date that not all larvae which remain in the larval stage throughout the winter develop into full females. There must, therefore, be still further factors which have an effect on caste determination. Such factors were investigated with *Myrmica rubra* L., a species closely related to *Myrmica ruginodis* (Brian, 1963; Brian and Hibble, 1964). Besides the presence of the queen, which has the same effect as a fourfold increase in the proportion of workers, blastogenic factors also have a part to play here, as with the wood ants. The more eggs a *Myrmica* queen lays within a specific time, the smaller these eggs will be. The smaller the eggs are, however, the narrower the chance of their developing into full females. In the natural environment, the year of the colony starts with the queen laying eggs after the winter rest period, at a temperature of about 20°C. Egg production then rises sharply, reaching a maximum after about three weeks. During the course of the year, the number of eggs laid gradually decreases until the end of the brood period. The size of the eggs alters correspondingly: the first eggs of the year are the largest. The egg size decreases during the first three weeks and then remains at a nearly constant level.

In addition to this, temperature, and the age of the queen affect the development of the larvae into full females or workers. Raising the temperature among the larvae from 22°C to 24°C in the first three stages of their development effects an eightfold increase in the proportion of workers produced from this brood. Apart from this, temperature clearly has an effect on the ability of the queen to lay eggs, as well as on the eggs she lays. The queen lays fewer eggs at 20°C than at 25°C. If the eggs laid at 20°C are then allowed to develop at 22°C, then the survival rate and the number of full females resulting is greater than at 20°C (Brian and Kelly, 1967). In selective experiments, *Myrmica* workers with their brood favoured temperatures of around, but not more than, 22°C. They thus selected temperatures which brought about the highest survival rate for the brood. These selective experiments did not give any indication that colonies well supplied with queens chose higher temperatures for the development of their larvae, so as to produce more workers and fewer full females (Brian, 1973c).

The age of *Myrmica rubra* queens played a part, insofar as young queens produce relatively more workers. The narrowness or otherwise of the brood chamber would also seem to have an influence on caste differentiation. The closer together workers and brood are in the nest, the fewer larvae are reared in equal egg production, and the less the growth of the larvae, which increases the proportion of workers in the developing brood (Kelly, 1973).

The *Myrmica* workers can intervene in the brood's development in a variety of ways. They can inhibit the growth of the larvae through the food which they give to them. In this case they feed them − with the same water content − less protein, and achieve by means of this 'meagre diet' a greater number of workers in the brood (Brian, 1973b). In addition, attacks on the part of the workers on large, potential queen larvae in spring work in such a way that the larvae pupate at a smaller size, thus becoming workers. This effect consists of bites from the workers directed at the head region of the larvae, and in the secretions of the mandible gland, which contain, amongst other things, poisonous substances (Brian, 1973a).

These various factors affecting the caste differentiation of *Myrmica rubra* work to a large extent, if not entirely, on account of the two antagonistic hormones, neotenin and ecdyson (Brian, 1974b). Here, as with all insects, neotenin is the juvenile hormone, which inhibits metamorphoses, whilst ecdyson not only facilitates the metamorphoses, but also accelerates their progress. Those influences on the larvae which lead to a greater production of ecdyson thus effect the occurrence of the metamorphoses at an earlier time when the larvae are smaller, thus also reducing the possibility that full females will be produced. The investigations of Brian (1974a), in which he treated *Myrmica rubra* larvae with synthetic ecdyson, showed that the effect of this treatment among the larger larvae, whose imaginal structures had begun to differentiate, was greater than at an earlier stage. However, treatment of the larvae with substances similar to juvenile hormones, delayed the metamorphoses and resulted in a larger number of full females.

6.4   The formation of the subcastes

Still very little is known about the origin of subcastes among ants, and the little we do know is concerned for the most part with the origin of the soldier caste. In the harvester ant genus *Pheidole,* three different larval stages can be discerned, differing mainly in the shape of the bristles and the formation of the mandibles. If one compares the larvae of future soldiers with those from which workers will later emerge, no difference is noticeable up to the third larval stage. The size and form of the larvae, as well as the composition of their haemolymph, are apparently completely uniform; differences only appear in the pre-pupal stage (Passera, 1974).

This is not to say, however, that the determining influences on the formation do not occur at an earlier stage. According to Wilson (1954), the variations in the nutrition of the larvae, the result of competition for food among the brood, is an important factor, and both Goetsch (1937) and Passera (1974) attach a great deal of importance to this. This thesis is supported by the experiments of Passera, in which part of a group of *Pheidole pallidula* was fed only with honey, while the other part of the group was fed with honey and animal protein. No soldiers developed from the group which had been fed only with honey, but none developed even from that part of the group which had also received protein. Temperature also plays a part in the emergence of *Pheidole* soldiers. At temperatures above 24°C, Passera obtained a percentage of soldiers of about 4% from the brood, whilst at temperatures below 24°C he still obtained workers, but no soldiers.

These factors, however, are not sufficient to explain the incidence of soldiers; what they do is lay down the necessary conditions for their formation, but this does not explain the constant proportion of soldiers found under natural conditions in the field. A regulating mechanism must be in effect here which holds the proportion of soldiers in the whole colony at a roughly constant level. This regulating mechanism could take effect through the nutrition of the larvae, or it could involve a direct influence on the part of the fully grown soldiers on the developing brood, as Gregg (1942) assumes. Gregg discovered in the case of *Pheidole* that in colonies consisting only of soldiers, only a few soldiers emerged from the brood, whereas in colonies where there were no soldiers, many soldiers would suddenly be produced from the brood. He concluded from his experiments that the soldiers of a *Pheidole* colony release a substance which affects the development of the larvae, by inhibiting the development of new soldiers.

As far as the origins of other subcastes are concerned, such as the origins of the minima workers of *Oecophylla,* we know only that they leave out a stage in their larval development, thus pupating prematurely and remaining smaller (Haskins, 1970).

7 **Regulation of sexuality**

It is decisively important for the continuation of the species that larval determination for full females is coordinated in time with the development of males in the colony; all the more so since in the majority of ant species the male lives only a relatively short time. It is therefore necessary to ensure that the emergence of full females and the development of males are sufficiently closely synchronized to enable mating. This means nothing more, however, than that a proportion of the eggs laid should remain unfertilized 'at the proper time'.

Among South American army ants, humidity is a decisive factor in the coordinated rearing of the sexually reproductive animals. Schneirla (1948) determined through observations in the field that when humidity in the atmosphere is reduced and it starts to become drier the sexual animal brood emerges, with a high proportion of males.

Among the army ant genus *Anomma* inhabiting the similar tropical climate of Africa, the regular emergence of males can, according to Raignier and von Boven (1955), be very simply explained. The sperm reserve of a queen is exhausted in this case before the end of one of the egg-laying periods, in which more than a million eggs can be laid. In this way, a proportion of the eggs will always remain unfertilized from which the males will develop to inseminate the *Anomma* queens.

Temperature plays an important part in the occurrence of the males in our European wood ant species, which, like the female sexual animals, develop from the first brood of the year. We know from observations of colonies of *Formica polyctena* that in the majority of cases only single sex broods of sexual animals are reared (Gösswald, 1951). The colonies of this species thus produce either only male or female sexual animals as a rule, apart from the workers which are bred later in the year. In those rare instances of a mixed brood of sexual animals, the males emerge before the females.

It is typical of colonies producing only males, either that they are small or that their nests are situated in particularly cool places. The reverse is true for colonies producing exclusively females. Either they nest in particularly warm places, or theirs are particularly large colonies. It is a common feature of all colonies that produce males in the spring that they are not able to raise the temperature in the nest after the winter rest period to the same extent as those colonies producing only females. This is the case both with small colonies and with those larger ones where the nest is situated in a particularly cool place.

In what way temperature affects the emergence of males with regard, for example, to the way in which a proportion of the eggs will remain unfertilized as a result of temperature influences, is show in the investigations of Gösswald and Bier (1957) into the wood ant *Formica polyctena.* According to these, the *Formica polyctena* queens begin to lay eggs at temperatures above 15°C. If the temperature is only just above 15°C, such as somewhere between 15 and 21°C, then 'sperm pump' activity begins later than egg-laying activity; this results in a proportion or all of the winter eggs being laid unfertilized. The term 'sperm pump', however, is used here only in a very broad sense. It describes an as yet unknown mechanism, whereby the sperm from the receptaculum seminis are transported into the oviduct, thus coming into contact with the eggs. We still do not know whether there really is a pumping process, or whether the sperm are activated by secretions from the appendicular gland and leave the receptaculum independently. All we know for certain is that the sperm do not leave the receptaculum until higher temperatures than are necessary for egg-laying are reached. As a result of this, most of the first eggs laid by the queen after the winter rest period are unfertilized and develop into males. This fact also explains the observation that in colonies with a mixed brood of sexual animals the males emerge before the females.

The synchronized breeding of males and females among *Formica polyctena* also has its origins — according to Schmidt, 1972 — in the fact that the larvae of male sexual animals, like the larvae of the female sexual animals, have to be fed by the 'spring workers' (workers hatched in the spring). Finally, another factor facilitating the synchronized development of the two types of sexual animal is the fact that the males only develop from winter eggs. Unfertilized summer eggs do not develop any further.

In the monogynous species *Myrmica sulcinodis,* sexual animals are probably only produced in old colonies, whilst in younger colonies only workers are produced. In other words, to the extent to which the colony has reached its maximum size and the queen 'ages', she correspondingly loses her influence on the caste determination of the brood. This leads to a corresponding increase in the number of female sexual animals produced. In this species, males probably result only from worker eggs in colonies where there is no longer a queen (Elmes, 1974).

Among the species *Leptothorax recedens,* which also belongs to the Myrmicines, the males also evidently emerge exclusively from the unfertilized eggs of workers. In colonies with queens the *Leptothorax* workers lay only trophic eggs, which serve for food and not for reproduction in the colony. This alters only at a specific time of the year, directly after the winter rest period. During this period, which lasts only a few weeks, the workers lay eggs which are able to develop, thus ensuring the necessary number of males for the year. Clearly, the physiological condition of the queens directly after the winter rest period involves a temporary loss of their inhibiting effect on their workers, thus making it possible for them to lay eggs which will develop (Dejean and Passera, 1974). The time at which laying activity begins in *Leptothorax* colonies is, by the way, strangely early, usually falling during winter itself. *Leptothorax* queens begin laying eggs at 7.5°C, and their workers at 10°C (Passera and Dejean, 1974).

8 Colony founding

After insemination, ant females frequently withdraw into a brood chamber and found their colony there without help from outside (independent colony founding). One cannot generalize where ants are concerned as to how frequently they have been inseminated before this. *Myrmica rubra,* for example, looks for a brood chamber directly after a single insemination (Brian and Brian, 1955), whilst we know from other species, either by comparing sperm quantities of males and females which have been inseminated, or by direct observation, that the females are inseminated a number of times. This is the case, for example, for the leaf-cutter ant species *Atta sexdens* (Kerr, 1962) and for the American slave-making ant *Polyergus lucidus* (Marlin, 1971).

In Western Siberia, Marikovsky observed that wingless females of *Formica rufa* which have already laid eggs as queens, will leave the nest at the time of the nuptial swarm, and may be re-inseminated by males. Mostly, however, they are hindered from leaving the nest, by the fact that the workers spray minute quantities of formic acid on the mouth parts of the queens.

Males may also mate successfully a number of times, as we know from the socially parasitic species of ant, *Anergates, Teleutomyrmex* and *Harpagoxenus.* The repeated mating of males appears to occur wherever the females' production of offspring is numerically small (Buschinger, 1970).

The founding of a new colony is, for the majority of ants, fraught with innumerable dangers. Even during the nuptial flight, the sexual animals may become prey to various predators, which lie in wait for the flying ants. When the females have survived the nuptial flight undamaged, the dangers that await them are not only further predators, but also the uncertain prospect of obtaining their own nest. Even those females which have finally found a nesting place are still far from secure. They may be attacked even in their brood chamber, or have to take account of the competition from already established colonies. The competition between established colonies and young colony-founding females emerges, for example, from the investigations of Brian (1955a, 1956b,c), who showed for the *Formica* and *Myrmica* genera that the number of colony-founding females and newly founded colonies decreases in proportion to the greater number of established colonies. Aggression between the workers of older colonies and young females of the same species also predominates among *Lasius niger* and *Lasius flavus* (Pontin, 1960a), among the wood ant *Formica rufa* (Marikovsky, 1962) and among the American harvester ant species *Pogonomyrmex barbatus*, *Pogonomyrmex rugosus* and *Pogonomyrmex maricopa* (Hölldobler, 1976a). In the majority of cases, colony-founding females in the vicinity of established colonies of the same species are discovered very quickly, and are attacked and killed by the workers of these colonies. Among the *Pogonomyrmex* species investigated by Hölldobler (1976a), on the other hand, as a rule such episodes proceed without bloodshed (*see below*). *Novomessor cockerelli*, also a harvester species, however, frequently captures swarming females and also males from *Pogonomyrmex* species and takes them back as food (Hölldobler, 1976a).

Whitcomb *et al.* (1973) investigated the various sources of danger for the fire ant *Solenopsis invicta* which has migrated into North America, the dangers of predators for the young females from the beginning of the nuptial swarm until the hatching of the first workers. According to these investigations, these are first, those from a series of dragonfly and bird species, which follow and capture the ants when flying. After the nuptial flight, the inseminated females are attacked on the ground, mainly by other ants – even their own species – as well as spiders, earwigs and beetles. Finally, in the brood chamber the females may still be attacked by ants (e.g. *Lasius neoniger*), earwigs and birds (quails). According to the calculations of Whitcomb *et al.* (1973), more than 99% of the female sexual animals fall prey to predators alone. According to Evans and Eberhard (1970), there are also sand wasps (Sphecids), for example, *Aphilanthops frigidus*, whose active periods coincide with the nuptial flight of *Formica fusca*, and which carries back exclusively young winged females as larva food.

In addition to independent colony founding by isolated females, there are among ants various forms of dependent colony founding which have come about. In this case, the young females are supported from the beginning of colony founding by workers of their own species. Among temporarily or permanently socially parasitic ants, the founding of the colony in nests is carried out by other species.

8.1   Independent colony founding

The most generalized form of independent colony founding is to be
found among the Myrmeciines and the Ponerines. The larger species of
the genus *Myrmecia,* such as *Myrmecia forficata* (Clark, 1934),
Myrmecia pyriformis (Tepper, 1882), *Myrmecia gulosa* (Frogatt, 1915,
after Wheeler, 1933) and *Myrmecia vindex* and *Myrmecia inquilina*
(Haskins and Haskins, 1964), take part in nuptial flights in which some
of the females swarm, but where others also fly singly. Among
Myrmecia regularis and *Myrmecia tarsata,* the wings are so reduced that
they are no longer fit for flying; they break off shortly after hatching
(Haskins and Haskins, 1955; Haskins, 1970). The females of this species,
like the females of the completely wingless species *Myrmecia aberrans,*
first walk around in the vicinity of the nest and are then trailed and
inseminated by males which are able to fly.

The inseminated *Myrmecia regularis* females (according to Wheeler,
1932 and 1933) hollow out a large flat cell under a stone and close
themselves off from the outside world. From the base of this chamber,
they dig a perpendicular tunnel, ending in a second, smaller chamber
(Fig. 8.1). This second chamber will be resorted to in case of danger
or too-great heat. The females do not, however, remain permanently
shut inside this two-chambered nest, which is the typical Myrmecine
kind. A number of the *Myrmecia regularis* females observed by Haskins
and Haskins (1955) left a gap permanently open in the nest. All the
females, however, left the nest every evening, fetched sweet foodstuffs
and returned immediately.

After about a month, during the Australian high summer in
November or December, the *Myrmecia* females lay their eggs. These
eggs do not hold together in clumps, as one observes among the 'higher
ants', but rather lie individually scattered in the brood chamber. As
soon as the larvae have hatched, the females also begin to hunt insects
and bring them in for food. According to the observations of Wheeler
(1932), these insects are the only form of food eaten by the developing
larvae. Haskins and Wheldon, however, were able to show in 1954 by
means of coloured sugar solution that the females in fact also give the
larvae liquid food from their crop.

Towards the end of their development, the larvae are covered with
earth, to which they are able to attach their threads for the construction
of their pupal cocoons. The fully developed workers are later released
from their cocoons by the females. The young *Myrmecia* workers leave

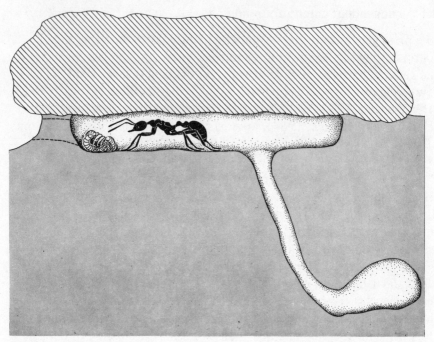

Fig. 8.1   *Myrmecia regularis* female in her brood chamber under a stone during colony founding (after Wheeler, 1933).

the nest as early as a few hours afterwards and bring in food. The young queen ceases to go out of the nest to the extent to which the young workers take over this task, and in the fully developed colony the queen probably stays in the nest at all times.

The independent colony founding of the primitive Ponerine genera occurs in the same basic pattern. This emerges, for example, from the experiments of Wheeler (1933) with *Amblyopone australis.*
Brachyponera lutea on the other hand, which counts among the more highly evolved Ponerines, is the only species of the genus known so far to display 'cloistral colony founding' (Haskins and Haskins, 1950), such as is otherwise typical of many Formicines.

Independent colony founding of ant species native to Germany has been described mainly by Gould (1747), Huber (1810), Eidmann (1926, 1928a, 1931, 1943), Meyer (1927), Stäger (1929), Goetsch and Käthner (1938), Hölldobler (1936, 1938b, 1950) Waloff (1957), Andrasfalvy (1961), Le Masne and Bonavita (1967), Buschinger (1968d) and others. According to these investigations, the most developed form of independent colony founding is the 'cloistral' form, which has been demonstrated for the species *Lasius niger, Lasius flavus, Formica fusca, Camponotus herculeanus, Camponotus ligniperda, Tapinoma erraticum, Tetramorium caespitum, Solenopsis fugax, Leptothorax unifasciatus* and *Prenelopis imparis.* The inseminated females withdraw, like the *Myrmica* females, into a brood chamber, which they render impervious on all sides and do not leave during the whole colony founding period. The freshly hatched workers of the young colony are the first to open the brood chamber and look for fresh food.

The particular advantage of this cloistral form of colony founding lies in the fact that the female, who never goes out to look for food, runs less danger of being found and captured as prey. On the other hand, this form of colony founding obliges the females under certain conditions, to wait a very long time until the first workers hatch and look for fresh food. The ability of colony founding females to go without food has been observed in the laboratory, and it was discovered that they were able to last well over a year without food.

Even more astounding than the frugality of the colony founding females is the fact that they lay eggs in these conditions and can only rear their first brood from their own food reserves. This, of course, can only be achieved in those species where the females are much larger than the workers (Fig. 8.2). The fat compounds and flight muscles of the females serve them as food reserves, the flight muscles no longer being used once the wings have broken off after the nuptial flight. The larger proportion of the eggs and the emerging larvae are destroyed by the queen and serve as food for the few larvae which continue to develop and eventually pupate. Among a number of species, the brood is fed to the larvae without first passing through the crop of the queen. It also occurs that the larvae feed themselves and eat their fellow larvae, as Eidmann (1929b) observed with *Formica fusca.*

The first workers of a young colony are particularly tiny, rather smaller than the workers which develop in a fully established colony. It would seem that this small body size is the result of their poor diet during the founding of the colony, but this has not been confirmed. Among *Lasius niger,* for example, the eggs of a colony founding female did not produce any larger workers when reared in an established colony. Among the slave-making ant *Harpagoxenus,* Buschinger and Winter (1975) also found that the first workers were only smaller when there were enough slaves and sufficient food available from the beginning. Among the Ponerine *Odontomachus,* the first workers are smaller on account of the fact that a larval stage has been omitted: in the founding colony they pupate earlier, after the third stage instead of the fourth (Ledoux, 1952).

In contrast with colony founding by isolated females, the joining

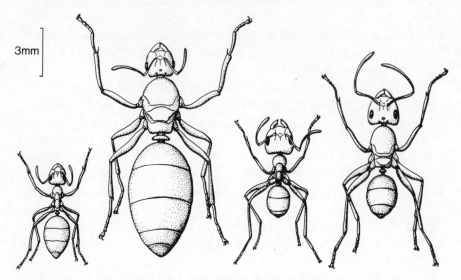

3mm

Fig. 8.2   The difference in size between workers and females in an independent colony founding species (*Lasius niger,* left) and in a socially parasitic colony founding species (*Lasius fuliginosus,* right) (after Eidmann, 1926, adapted).

together of several females and the founding of a joint colony also occurs. Such pleometrosis is frequently found among the yellow *Lasius flavus,* which lives underground (Eidmann, 1926) and *Lasius niger* (Eidmann, 1926; Donisthorpe, 1927), and less frequently among other species, such as the fire ant *Solenopsis invicta* (Markin *et al.,* 1972). The advantages of pleometrosis for *Lasius flavus* emerge from the investigations of Waloff (1957): among groups of colony founding females, the mortality rate of the females was lower, and the rearing of the first workers occurred more quickly than with single females. The advantages of the joint founding of a colony have also been shown with *Camponotus vagus:* according to Stumper (1962), two females founding a joint colony showed greater egg-laying activity and more care of the brood than individual females. Under natural conditions, however, *Camponotus vagus* colonies are generally founded by individual females (Benois, 1972a).

Since only one queen capable of reproduction lives in an established *Lasius flavus* colony, the transition from pleometrosis to monogyny has to occur during the founding of the colony by groups of females. This usually takes place as a division of the young colony corresponding to the number of colony-founding females. According to the observations of Wasmann (1910a), this occurs early on, at the time when the first larvae form. According to the observations of Waloff (1957), however, the division of the young colony occurs later, usually when the first larvae pupate.

Among *Lasius niger,* a fight between the colony-founding females decides which of them will remain as the single queen of the colony. Among carpenter ants also, monogyny is created by means of a fight among the females (Hölldobler, 1962). The carpenter ants native to our part of the world, in particular *Camponotus herculeanus,* inhabit the insides of living trees, so that one nest may include many trees (*see below*). Under these conditions, it can happen that the queens never meet, so that several queens live in the colony. Among the Myrmecine species *Crematogaster striatula,* on the other hand, as with established *Lasius* colonies, the workers ensure that there is always only one queen in the colony. Here, therefore, it is the workers who maintain monogyny, by killing any other females capable of reproducing (Soulié, 1964).

Other indigenous ant species, such as *Ponera* species in particular, but also the majority of *Myrmica* species, follow the same pattern as the Myrmeciines when founding their colonies. The food reserves of the females in this case are not so large as to enable them to rear the first brood without an additional food supply. It has thus been demonstrated that they fetch food during the founding of the colony, as Hölldobler (1938b) has shown for *Myrmica rubra* and *Myrmica lobicornis.*

Myrmica rubra females are often inseminated actually in the nest. They shut themselves afterwards inside a chamber, singly as well as in groups of several females, but they leave the chamber at regular intervals in order to search for food. Significantly, the laying of eggs in the spring of the following year only seems to be possible where the females have previously fetched in food. With *Manica rubida* also, another of our indigenous Myrmicine ants, independent colony founding is not possible unless the females go out repeatedly in search of food. This has been demonstrated by Le Masne and Bonavita (1967), contradicting the converse supposition of Eidmann (1926).

Specialized aspects of independent colony founding are shown chiefly by those species which limit themselves to a specific diet. This is the case, for example, for the South American leaf-cutter ants of the *Atta* genus, which cultivate a specific fungus on plant material, and probably live exclusively from the products of this fungus (*see below*). The *Atta sexdens* females thus take a few fragments of mycelia from their fungal garden in their 'infrabuccal sack' at the rear end of the mouth orifice (*see* Fig. 1.4) with them on their nuptial flight (von Ihering, 1898). The *Atta* females differ with this purely vegetable culture deposit from a number of beetles and termites, which transport the spores of their fungus (Weber, 1972b).

The process of independent colony founding among leaf-cutter ants was first investigated in greater detail by Huber (1905) with *Atta sexdens.* According to these investigations, the *Atta* females bury themselves into a brood chamber and attend firstly to the growth of their fungus. The first eggs are laid about three days after the females have dug themselves in, when the fungal growth is already showing delicate threads spreading out in all directions. In the following 10-12 days, the female lays about ten eggs a day, which are at first stored in a place separate from the fungal garden. After 8-14 days, however, when the fungal texture has reached a diameter of 1 cm, the eggs are deposited on the fungus. Of these eggs, however, nowhere near all of them come to develop. Over 90% of them are consumed as food by the females, or fed to the larvae when the female presses the eggs on to the mouthparts of the larvae, allowing them to suck them. These eggs form the only diet for the larvae during the founding of the colony: the fungal material is not touched during this time. Only the workers that hatch from between the second and third month after the beginning of colony founding (Fig. 8.3) begin to consume fungal material. Bazire-Benazet discovered in 1957 that the colony founding *Atta* females lay two different types of eggs, those which are able to develop, and particularly large eggs, which serve only as food. These oversized eggs result from the fusing of several eggs and are described as 'omelettes'.

One pleometrotic variation of this colony founding is shown by the leaf-cutter ant *Acromyrmex lundi* which occurs in the Argentine, where several females join together for the founding of a colony and establish a joint fungus garden (Weber, 1972b).

The species *Acropyga paramaribensis* living in Sumatra has similarly specialized in a particular diet, living exclusively from the extracts of the scale insect *Pseudorhizoecus coffea*: the swarming *Acropyga* females

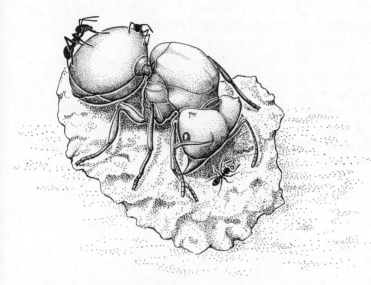

Fig. 8.3   Young colony of a leaf-cutter ant (*Atta*) with queen, first workers and fungus garden (after Weber, 1972, adapted).

take an inseminated insect between their mandibles and place them after their nuptial flight at the root of a coffee plant. Only after this point does independent colony founding begin, when the reproducing insects are looking for food (Bünzli, 1935).

8.2   Dependent colony founding

In independent colony founding, the young queen is helped from the beginning by workers. This form of colony founding is presumed to exist for a number of species of *Carebara* occurring in tropical Africa. These species live mostly as thieving ants (*see below*) in termite structures. The striking feature of the *Carebara* genus is the extreme difference in size between the tiny yellow workers and the much larger sexual animals. According to Arnold (Wheeler, 1923), it should as a result of this difference in size be difficult for the females to feed their brood, so much smaller than themselves, during the founding of the colony. Wheeler in fact found — as Arnold had done before him — that the young females are accompanied by workers on their nuptial flight, the workers biting into the dense hair on the tarsi of the females' legs (Fig. 8.4). Arnold assumed that these workers assist the young queens during colony founding, chiefly by feeding the first brood. We still do not know the extent to which the *Carebara vidua* females actually do depend on the help of the few workers which are transported with them, since no-one has yet been able to observe the colony founding of this species. It has been demonstrated for *Carebara lignata,* however, which was investigated by Lowe (1948) in Malaysia, that under natural conditions the females are able to found their colonies even without the help of workers. The females of this species take no workers with them on their nuptial flight, and they are, in comparison with their workers, if anything even larger than the *Carebara vidua* females. *Carebara lignata* is 19 mm long, and has workers 1.5 mm long, the length of *Carebara vidua* queens is 24 mm, whilst that of their workers is 2 mm.

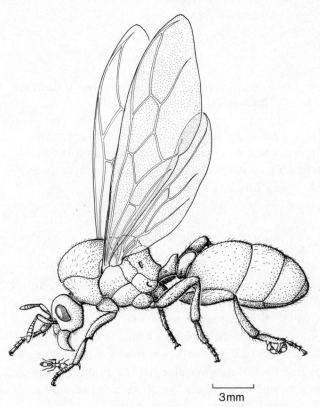

3mm

Fig. 8.4   *Carebara vidua* female, which takes workers (hanging from the legs) on its nuptial flight (after Wheeler, 1923, adapted).

Dependent colony founding has been described with more certainty with *Odontomachus haematodes,* which has been investigated by Colombel (1970a, 1970b, 1971b,c,d, 1972a,b,c). The *Odontomachus* females are also able to found relatively small colonies independently with 300 workers. Under natural conditions, however, it frequently occurs that the colony-founding females are to be found with workers of their own species, which join with the females to form a colony, rearing the first brood together. The attraction of the young colony for further workers is reduced when the ratio between the workers and the queen reaches about 85:1, which Colombel also found to be the case in older colonies. This form of dependent colony founding is also called 'plugging', and had already been described by Marlin (1968) in connection with the American slave-making ant *Polyergus lucidus.*

The dependent colony founding of *Aphaenogaster senilis,* which was investigated by Ledoux (1971, 1973), takes place in a different manner. In this species, new colonies emerge by existing colonies dividing. Without the emergence of another queen to succeed her, the old queen leaves the nest with somewhere between 150 to 250 workers, and re-colonizes in the vicinity. In the old colony, where the influence of the queen is now lacking, female sexual animals develop from a portion of the brood, of which one as an inseminated female will take over as successor to the old queen. She will be inseminated by a male which has emerged from the brood of the remaining workers.

Among the African Dorylines of the *Anomma* genus also, new colonies emerge through division (Raignier, 1959, 1972), which of course requires the presence of a male brood. Male sexual animals can, in fact, develop all year round, but they are particularly prevalent during the latter third of the dry season. But not every emergence of a male brood leads to a division, and not every division necessarily leads to a new founding of a colony. The founding of a new colony is only possible when a new queen has previously emerged. The processes leading to the emergence of young queens and to the founding of a colony are as follows: the male *Anomma* brood receives much more food than the group of worker larvae, which is more numerous than the males. A fertilized egg — or even several — developing in the colony at this time will partake of this richer diet and, affected by this, will develop into a female sexual animal, hatching some 14 days before the males. Workers gather round this new queen and the male pupae, gradually detaching themselves from the rest of the colony. After new workers have hatched, the old queen withdraws with her contingent of workers, leaving the new queen and her contingent behind. The new queen is inseminated after the males have hatched and begins to lay eggs.

The Ponerine genus *Rhyriodoponera* includes species whose workers may be inseminated, thus serving the colony as functional queens. Whelden (1957) counted 8% of workers as fertilized in one *Rhytidoponera convexa* colony, 2.3% with *Rhytidoponera inornata* and 5.4% with *Rhytidoponera metallica* (Whelden, 1957, 1960). Full females were in fact also found among *Rhytidoponera metallica*, but these were very weak and not sufficiently developed for the tasks of independent colony founding. Among these species, new colonies emerge by division, a proportion of the fertilized and non-fertilized females withdrawing and making themselves into an independent colony (Haskins, 1970).

Re-colonization by division of old colonies can also be found among our indigenous polygynous species, in which inseminated females are taken into the mother colony after the nuptial flight. This is the case, for example, for the wood ants *Formica polyctena, Formica lugubris, Formica aquilonia* and also *Formica (Coptoformica) exsecta* (Gösswald and Schmidt, 1959), for the Pharaoh ant *Monomorium pharaonis* (Peacock *et al.,* 1955), for a number of *Crematogaster* species (Soulié, 1962) and for *Myrmica rubra* (Brian and Brian, 1955). Among *Myrmica ruginodis* there are, as Brian and Brian have shown, two forms of the female, differing in size, micro- and macrogynes. Whilst females of the macrogyne form found new colonies after the nuptial flight, the microgynes return to the mother colony.

The division of polygynous wood ant colonies had already been described by Brun (1910) and was later described by many other authors. In this form of division, a proportion of the queens withdraw from the nest with workers and brood, and set up a branch colony close by. This formation of branch nests by the wood ants differs considerably from the swarming of the honey bee. Whereas the swarming bees leave the colony once and for all with their still unfertilized queen, dissolving all links with the old colony, the removal of the ants and the formation of a branch colony is a process which may be spread over several days or weeks. During this time, workers run busily to and fro between the mother nest and the new nest site, transporting brood and nest mates into the new domicile. Transportation occasionally occurs in the opposite direction also.

At the beginning of such a removal, it is not possible to say with any certainty what its outcome will be. It may be that the new branch nest will be given up, and the entire colony relocate itself in the old nest. It is also possible, however, for the mother nest to completely empty itself and for the entire colony to remove into the new nest. It can in addition also happen that the ants only change nests for a season, for example for the winter or the summer only, and then return to the old nest ('seasonal nest change'), which occurs particularly frequently among *Formica pratensis.* Finally, it also occurs that the colony splits up. In this case the connections among the colonies are not usually broken. Mother and daughter colonies may continue to maintain contact, and later exchange workers and brood. Those colonies resulting from division which also practice exchange are termed 'polydomous colony associations'. They can be seen in the species *Formica polyctena, Formica aquilonia, Formica lugubris* and *Formica (Coptoformica) exsecta* (Gösswald and Schmidt, 1959; Dobrzanska, 1973). Such colony associations may occasionally contain a large number of nests and occupy whole areas of forest. In 1877 Mac Cook counted in the North American Alleghanies 1600 branch nests of the species *Formica exsectoides,* which together formed a polydomous association of colonies.

A proportion of wood ant females are inseminated in the nest; these females miss the nuptial flight and the return into the old nest or being accepted into an alien nest of the same species (Escherich, 1917; Moczaŕ, 1972). Among the house ant species *Iridomyrmex humilis* also originating in the Argentine the females are inseminated in the nest, directly after hatching in fact. The number of functional queens in the colony nevertheless remains the same, as the workers kill a corresponding proportion of their queens in January or February of the following year (Markin, 1970).

Among *Crematogaster scutellaris, Crematogaster auberti, Crematogaster sordidula, Crematogaster skouensis* and *Crematogaster vandeli,* Soulié (1962) observed branch nest formation, which occurred by a proportion of the workers withdrawing from the mother nest without queens, and later adopting a swarming female. A necessary precondition for an adoption, however, is that the female should be inseminated; non-inseminated queens are either frightened away or killed. The means by which the workers distinguish the inseminated from the non-inseminated females remains their secret. If a group of withdrawn workers does not succeed in adopting a female, then the queenless branch colony expires with the last of its workers. Apart from the formation of branch nests, *Crematogaster* colonies may also be founded by isolated females. This occurs relatively frequently among *Crematogaster scutellaris,* but very seldom among *Crematogaster auberti* and *Crematogaster skouensis.*

Crematogaster impressa from the savanna lands of Africa also founds new colonies with isolated females and by the removal of groups of workers from the old colony (Delage-Darchen, 1974). These workers do not obtain their queen by adopting a swarming female, however, but by rearing females themselves. They probably take fertilized eggs with them from the mother nest for this purpose, which develop into females in their care. These females are later inseminated by males which have grown from the eggs of the migrant workers.

8.3 Temporary social parasitism

Temporarily socially parasitic ants are dependent on the help of other ants for a certain period. In all known cases, this period is the time when the colony is being founded. The females of socially parasitic ants are not in a position to found their colonies independently, usually not even with the help of workers of their own species. They invade the colonies of other species, usually killing the host queen and allowing their brood to be reared by the workers of the other species. The host colony, robbed of its queen, gradually dies out, whilst the number of social parasites increases. With the death of the last host worker, an independently existing colony of social parasites finally emerges from the mixed colony.

Among the mostly polygynous species of our indigenous wood ants, new colonies emerge either by division or by socially parasitic means. This is the case, for example, for the species *Formica pratensis, Formica aquilonia, Formica lugubris, Formica truncorum, Formica uralensis* and probably also for all *Coptoformica* species, for example, *Formica (Coptoformica) exsecta* (Kutter, 1969). With *Formica polyctena,* on the other hand, new colonies are formed only by division, whilst the monogynous form of *Formica rufa* founds colonies by socially parasitic means only.

In the socially parasitic colony founding of our indigenous species, the females which have not found their way back into the nest go, after ridding themselves of their wings, to a nest of their helper ants. As far as their helper ants are concerned, these are of the Serviformica group, that is *Formica picea, Formica fusca, Formica lemani, Formica cunicularia, Formica rufibarbis* and *Formica cinerea,* which found their own colonies independently. The invasion of the wood ant females into the helper ant nest is full of risks, and many females die in the attempt. Only relatively few females succeed in being accepted — those who because of their bodily structure, and further because of their behaviour, are particularly suited to undertakings of this kind. They are neither very strong nor skilful in camouflage. Those, however, that succeed in being adopted, bite off the head of the host queen (Wasmann, 1909).

Bothriomyrmex decapitans invades the nests of its helper ant with greater ease (Santschi, 1920): the inseminated *Bothriomyrmex* females allow themselves to be seized without resistance and dragged into the nest by the workers which run out for them. Once inside, they quickly seek refuge with the brood, or on the back of the queen, and are here protected from the attack of the workers. On the queen's back, which all parasitic females eventually seek out, they set to sawing off the head of the queen — often in the course of several hours. Afterwards they are adopted by the workers, since they have in the meantime acquired the scent of the nest. They lay eggs, allow the brood to be reared by the helper ants and eventually become queens of pure *Bothriomyrmex* colonies.

We also know of socially parasitic colony founding in a number of *Lasius* species. *Lasius niger, Lasius alienus, Lasius emerginatus* and *Lasius flavus* found their colonies independently with isolated females *Lasius umbratus,* on the other hand, *Lasius rabaudi, Lasius bicornis, Lasius carniolicus, Lasius reginae* and *Lasius fuliginosus* are temporary social parasites, already know for the fact that their females are not much larger than the workers (Fig. 8.2). The females clearly do not have sufficient reserve substances to found a colony independently, and since *Lasius* species are usually monogynous, the way is also barred for the female to be accepted in a colony of the same species.

The colony founding of *Lasius umbratus* had already been described by Crawley (1909) and later by Gösswald (1938) and Hölldobler (1953). According to these, the inseminated *umbratus* does not seek out her helper ants immediately, but firstly seeks a place to hide herself and there spend the winter. My own observations and information from Buschinger and Maschwitz, however, show that the young *umbratus* females also seek out the nests of their helper ants directly after the nuptial flight. Faced with a suitable nest, they first move around it without invading immediately; rather they attack one of the helper ant workers, drag it into a hollow or on to a blade of grass, and there bite it to death. Only then, after the 'killer instinct' has worn off, do they enter the nest of their helper ants. Neither the *niger* nor the *alienus* workers are very hospitable to begin with, but their hostile dragging at the feelers and legs of the parasite female gradually decrease, finally giving way to a definite growing interest in her, and she is then licked by the workers, in particular at the tip of the rear section of her body. A glandular secretion from the *umbratus* female clearly plays a part here, which eventually results in the *umbratus* queen being favoured over their own *niger* queen. The *niger* queen meanwhile becomes increasingly isolated and is finally killed by her own workers. According to the observations of Hölldobler, it may also occur that *umbratus* females, not having killed a worker before invading the helper ant nest, will then kill the host queens themselves.

The significance of the *umbratus* female's behaviour in killing a worker before invading the nest of the helper ants has not yet been explained. It could possibly represent a behavioural relic, reminiscent of times when the females would have had to invade the helper ant nest in face of resistance from the workers. Wilson (1955) did not find this behaviour pattern among the American *Lasius umbratus,* and therefore assumed that it could be peculiar to *Lasius rabaudi*, which occurs only in central Europe and which was previously indistinguishable from *Lasius umbratus.* The observations of the author, however, show that this killer behaviour in fact does also occur after this new species separation.

Just as puzzling as the behaviour of the *umbratus* females is that of the helper ant workers, in neglecting and finally killing their own queen in favour of the invading *umbratus* female. According to Buschinger (1970), however, it is conceivable that the parasitic *umbratus* female uses the monogynous regulation of her host colony to her own advantage, by being able through chemical means to be more attractive as a queen than the host queen herself. The workers, who see to the regulation of monogyny among *Lasius,* then decide in favour of the most attractive queen and kill their own queen. The colony founding of other *Lasius* species, on the other hand, which are closely related to *Lasius umbratus* such as *Lasius bicornis,* remains completely unknown.

The females of the golden yellow *Lasius reginae* first discovered in 1967 by Faber in lower Austria, found their colonies with the help of *Lasius alienus.* The *reginae* females are, in fact, substantially smaller than the *alienus* queens and are actually smaller than their own workers, but they nevertheless manage, after invading a helper ant colony, to turn the legitimate queen over on her back and bore through the unprotected spot between the head and the chest of the queen with their sharp mandibles (Fig. 8.5). After this the female begins to lay eggs,

Fig. 8.5 Female of the temporarily socially parasitic species *Lasius reginae,* having swung a queen of the host species over on to her back, kills her with one bite (after Faber, 1967).

during which the rear section of her body swells markedly because of the large number of developing eggs inside it (physogastry).

Lasius carniolicus, which is very similar to *Lasius reginae,* also belongs – as we have since discovered from the observations of Schmidt (according to Buschinger, pers. inf.) – to the temporary social parasites. Like *Lasius reginae, Lasius carniolicus* also have unusually broad heads with powerful glands whose secretions smell strongly of lemon. We know from *Lasius carniolicus* that this secretion is particularly rich in citronella and geranylcitronella (Bergström and Löfqvist, 1970). The *carniolicus* females are also smaller than their workers, and similarly possess sharp, dagger-like mandibles. They found their colonies with the help of the pale yellow *Lasius flavus* species.

The jet ant, *Lasius fuliginosus,* is the largest of our indigenous *Lasius* species. This species also belongs to the temporary social parasites, at least to the potential parasites. Their females are also able to be accepted back into the mother nest and form branch colonies. *Lasius umbratus* serves as the helper ant during socially parasitic colony founding. Although *Lasius fuliginosus* is a relatively prolific species, we still do not know how the *fuliginosus* females eliminate the *umbratus* queens.

In contrast with the *Lasius* species mentioned, the blood-red ant

Formica sanguinea can only be described as a temporary social parasite
in a rather benevolent sense. In the majority of cases, this species lives
together permanently with its helper ants, which would stamp them as
permanent social parasites, were it not for the fact that they are quite
capable of living without helper ants. In the case of a species living
together permanently with its helper ants, we speak of slave-making, or
'dulosis'. For dulotic *sanguinea* colonies the problem presents itself — as
it does for all slave-making ants — of obtaining the slave contingent,
since the queen of the slave species has previously been killed. The
slave-makers solve this problem by making regular 'war expeditions' to
colonies of the slave species, and there stealing pupae.

In the case of *Formica sanguinea* as the only indigenous potentially
dulotic ant species, the question that most needs to be asked is, of what
use are the slaves to them? This question, however, cannot be answered
satisfactorily. Dobrzanski (1961) assumes that it is not the lack of
helper ants, but rather the search for prey that leads the predatory ants
to their slave species (*Formica fusca* and *Formica rufibarbis*). The very
unorganized progress of the predatory expeditions compared with those
of the obligatory slave-makers speaks in support of this idea, as well as
the inability of the *sanguinea* workers to distinguish between the pupae
of their slave species and the pupae of other species. The greater
proportion of the pupae brought in by the *sanguinea* workers is in fact
eaten (Forel, 1874; Wasmann, 1891; Ghigi, 1951). The slaves kept in
the *sanguinea* colonies formed, according to Dobrzanski, only from
those pupae which could not be consumed on account of an excess
demand for prey. Kutter (1969) on the other hand, points to the fact
that only pupae, and never eggs, larvae or sexual animals are stolen from
the colonies of the slave species, and that the stealing expeditions only
ever occur at the time when the pupae are maturing.

A variety of possibilities are at the disposal of the predatory ants for
colony founding, as Wasmann (1910a) enumerated. New colonies may
emerge among *Formica sanguinea* by the inseminated females being
adopted either into the mother nest, or into colonies of the same
species; branch colonies may later form from such colonies. *Formica
sanguinea* colonies emerge in a socially parasitic manner by the
sanguinea inseminated females allowing themselves to be adopted by
Formica fusca colonies and killing the *fusca* queen there if necessary, if
the colony was not queenless already. The *sanguinea* females may,
however, place themselves in a brood chamber together with
inseminated *fusca* queens, allowing their own brood to be reared by the

fusca females, which are capable of founding a colony independently. Later, when the first workers have hatched, the *fusca* females are killed by the *sanguinea* predator ants. This form of colony founding, in which females of different species join together, is called alliance. *Sanguinea* colonies may furthermore result when inseminated females invade a *fusca* nest, carry off some of the pupae and allow them to hatch in a hiding place. The hatched workers later become the helper ants in the founding of a *sanguinea* colony. Finally, the *sanguinea* females may also take part in the stealing of pupae, but then remain behind in the plundered nest and found their own nest there with the pupae which have been left behind.

9 **Permanent social parasitism**

Permanently socially parasitic ants are dependent on the help of ants for a specific time and unable to live without this help. This group of permanently socially parasitic ants is quite heterogeneous, in particular in the means by which they ensure their stock of helper ants. The dulotic social parasites, or slave-makers, obtain their contingents of slaves on a regular basis by means of predatory expeditions. Among the non-dulotic social parasites there are those which, like the slave-makers and the temporarily socially parasitic ants, kill the queens of their host species. They pay for this by only being able to produce offspring for a few brood periods, after which they must then die out along with their hosts. Other social parasites allow their host queen to live, so that workers of the host species, and in some cases also sexual animals, will continue to be produced.

9.1   Social parasitism with dulosis

The best known of the dulotic social parasites is *Polyergus rufescens,* the Amazon ant. This species belongs to the Formicines and lives in dependence on *Serviformica* species. This dependence of the Amazon ants on their slaves has reached such an extent that without their help they would be unable to obtain any food for themselves. Even where there is the most rich food source, these slave-makers are obliged to go hungry if they are not fed by their helper ants. In contrast with *Formica sanguinea* colonies, as a result of this, the proportion of slaves among *Polyergus* does not decrease with increasing population density, but increases at a rate corresponding to it. Among *Polyergus rufescens,* the slaves in the colonies are something like five times more numerous than the slave-makers.

Their narrow sickle-shaped mandibles are the typical characteristic of the Amazon ants — only under the microscope is one able to make out the traces of hooks on them. These mandibles are no longer suitable for use in caring for the brood and building tasks; they are instead specialized tools for fighting and for the transportation of pupae. The Amazon workers spend most of their time in the nest cleaning their bodies, whilst the slave workers fetch food into the nest, see to the care of the brood and carry out building tasks. In summer, however, in the months of July and August, the Polyergus workers are also active. This is the time of the predatory expeditions, in which the Amazon ants seek out *Serviformica* nests in close formation almost daily. These processions are so impressive that they were mentioned by Latreille as early as 1805, and it is to him that the Amazon ants owe their name 'Polyergus' (initiating many wars).

Descriptions about these predatory expeditions generally agree that the *Polyergus* workers leave the mother nest quickly, proceeding in closed formation. Having arrived at the helper ant colony, they storm it as a group in commando fashion, emerging laden with pupae after a few minutes only to return to their own nest. The return to the *Polyergus* nest proceeds in a swift and haphazard fashion. Workers of the pillaged nest following the Amazon ants pay for their eagerness with their lives in many cases: they are easily pierced by the mandibles, which prove to be fearsome weapons.

The method of orientation of the Amazon ants during their predatory expeditions was disputed for a long time. Forel (1874) and Wasmann (1934) noticed a few Amazon workers a short time before the expeditions positioned in the vicinity of the attacked nest, and assumed that these workers served as scouts which would later lead the expedition. Dobrzanska and Dobrzanski (1960) were able to show, however, that the individual ants at the head of the expedition change continuously, and that the head group can move quite some distance away from the rest without the expedition altering course. They concluded from this that the scouts observed by Forel and Wasmann served only as 'activists', which would initiate the expedition by recruiting. They also concluded that the course actually adopted by the expedition after this recruitment occurred by chance. Köhler (1966) concluded the same from his experiments, in which he diverted the expedition by means of scattered pupae, and determined by comparison with control expeditions not influenced in this way that their yield of returned pupae remained the same.

The closely related American species *Polyergus lucidus* studied by Marlin (1969) seek out a *Serviformica* nest suitable to be attacked, and lay a pheromone trail between the *Polyergus* and the slave nests. This trail is intensified by further scouts, and then serves the predatory expedition as an aid to orientation.

Among the striking black- and red-coloured slave-making species *Rossomyrmex proformicarum* discovered by Arnoldi (1928, 1932) in Russia, the workers are carried to attack the colonies of the slave ants. These interesting desert ants, researched in more detail later by Marikovsky (1974), live in the stone, salt and lime deserts of Russia. Its helper species is the 'honey ant', *Proformica epinotalis,* a species which has special storer ants (replete ants). A *Rossomyrmex* colony contains between 23-200 *Rossomyrmex* individuals and 65-650 slaves, of which the smaller ones procure food, whilst the larger, storer animals are available to slaves as well as the slave-makers as living honey jars. Activity in the mixed colony begins promptly in the spring. Later in the year, during the great heat of July and August, the entrances to the nest are closed and the rest period beginning at this time lasts throughout the winter.

A thorough inspection of the area surrounding the helper ant nest to be attacked precedes the expeditions themselves. The small number of initiators of the predatory expedition then go back to the mother nest, there seizing one of their nest mates and carrying it to the helper ant nest which has just been the object of their reconnaissance. The animals which have been carried, after looking around, then proceed to transport further *Rossomyrmex* workers. This continues in geometric sequence until sufficient workers for attack have been assembled.

The preparations for the attack are not usually missed by the helper ants, who proceed to try to protect their nest. As soon as the first *Rossomyrmex* scouts appear, they close the entrances to their nest with small particles of earth, and even the corpulent storer ants make their contribution towards defence. They venture out of the lower chambers in the nest and stick the distended rear sections of their bodies like living corks in the entrance holes. The *Rossomyrmex* workers which eventually attack firstly remove these obstructions in the nest entrances: they tear off the heads of the replete ants, and then with great difficulty, remove the bodies of the storer animals from the openings. After the nest entrances are finally open, the actual attack begins, during which, astonishingly enough, the *Proformica* workers offer no resistance. They do not stand in the way of the invaders, but avoid them as much as possible, as a result of which, apart from the beheaded replete ants, no lives are lost during the attack.

The predator ants then seek out in the raided colony not only pupae,

but also take larvae and eggs with them, and even the bodies of the
replete ants are transported away.

It has been reported of other slave-making species, moreover, that not
only the helper ant brood was taken away for slaves, but even fully
developed animals. This phenomenon has been observed among
Formica pressilabris (Kutter, 1956) and *Strongylognathus alpinus*
(Kutter, 1921, 1969), and is called eudulosis. Such typical slave-making
expeditions as are represented by eudulosis, may perhaps be attributed
to causes which have nothing to do with slave predation, however. Since
the first indication by Elton (1932) that wood ants maintain a territory
over individuals of the same as well as other species, this phenomenon
has been discovered to exist in numerous other ant species: we can
assume that the majority of ants possess a territory. If the populations
of a species are then brought so close together that they are unable to
maintain possession of territory, a merging reaction comes into play,
whereby the populations merge together and then eliminate the sexual
animals of one of the two colonies involved (Buschinger, pers inf.;
Soulié, 1964). It is thus possible that eudulosis also in reality represents
a merging reaction, or a mixture of this and slave predation. This is
supported by experiments with *Harpagoxenus* ants, which display
colony-merging similar to eudulosis in relation to their slave species
when in small territories; in larger and more structured areas, on the
other hand, in which the helper ant colonies as well as those of the
Harpagoxenus may possess territories, the typical predatory expeditions
occur (Winter and Buschinger, pers. inf.).

Strongylognathus is a slave-making genus of the Myrmicine subfamily.
We have long been acquainted with about 20 *Strongylognathus* species,
all of which possess sabre-shaped mandibles in the manner of the
Amazon ants. The workers of *Tetramorium caespitum* serve
Strongylognathus as helper ants: the former is the so-called grass ant, a
prolific species found at heights of over 2000 m as well as on lower
levels. *Strongylognathus alpinus* has so far been found only in Wallis in
Switzerland, at heights above 1500 m. The occurrences known are at
Zermatt, Binntal, Lötschental and in Val d'Anniviers (Buschinger,
1971c). The predatory expeditions of *Strongylognathus alpinus* were
observed by Kutter (1921, 1969), who reported the following: the
alpinus predatory expeditions tend to take place during the night.
During these raids, the *Strongylognathus* workers and their slaves bore
underground passages until they reach the nest area of neighbouring
Tetramorium colonies and there try to force their way into the brood

chambers against resistance from the *Tetramorium* workers present. They usually succeed in their expeditions owing to their special fighting methods. Whilst the slaves brought on the expedition engage in heavy fighting with the workers of the invaded colony, *Strongylognathus* seizes its opponent from behind, sticking doggedly to its head. Sooner or later, the resistance of the attacked ants breaks down and the plundering of the colony begins, finally developing into a removal of the contents. The slaves brought by the *Strongylognathus* ants firstly clear out the brood chambers and then trail the workers of the invaded colony back into the *Strongylognathus* nest. Only the queen and any other existing sexual animals of the invaded nest are killed: all the remaining living contents are taken back and incorporated into their own colony. *Strongylognathus huberi,* which was detected by Forel (1874) in the main valley of the Rhone, is probably also an active slave predator. This species, however, was not found again for a long time after Forel's first discovery, until it was rediscovered almost a hundred years later by Baroni-Urbani in Switzerland, northern Italy and southern France (Baroni-Urbani, 1969).

   Strongylognathus testaceus is smaller than *Strongylognathus alpinus* and *Strongylognathus huberi* and is no longer a slave predator according to Kutter (1969). This is to say that a mixed *Strongylognathus* colony always contains a reproductive queen of the helper ant species (*Tetramorium caespitum*) and the parasite species and whose offspring workers in the case of *Tetramorium* are far in excess of those of *Strongylognathus*. Whilst, however, the *Strongylognathus* queen produces a sexual brood of considerable size, only workers result among the broods of the helper ants. In this way the helper ants no longer serve to maintain their own species, since they now only serve to maintain their parasite species. The assistance of slaves is also necessary in the colony founding of *Strongylognathus testaceus,* however still nothing is known about this.

   The *Harpagoxenus sublaevis* prevalent in central Europe (Buschinger, 1971d) lives together with slaves of the species *Leptothorax acervorum, Leptothorax muscorum* and *Leptothorax gredleri* (Buschinger, 1966b,c, 1971d). *Harpagoxenus* is in fact dependent on its helper ants, but is nevertheless able to fetch food independently. The presence of winged and wingless females was originally taken by Viehmeyer as an indication of different strains in the species, whereas Buschinger (1966a) showed that winged females, which occur only rarely, may also result from the eggs of wingless females. According to Buschinger (1975) a dominant allele E prevents wing formation, whilst an allele e only permits a homozygotic determination for the larvae to develop into winged females.

The predatory expeditions of *Harpagoxenus sublaevis* are much less spectacular than those of *Polyergus*. In the field, they take place some time between June and August (Buschinger, 1967, 1968a), that is during the time when one would expect to find pupae in the colonies of the helper ants. Buschinger investigated these expeditions (1968a) in the laboratory, and observed the following process: a *Harpagoxenus* worker which has discovered a helper ant nest in its role as scout fetches reinforcements in the manner already described, who then force their way violently into the *Leptothorax* nest. After heavy fighting, the workers of the attacked colony eventually flee; otherwise, their legs and feelers will be cut off as has already happened to their co-workers before them. Afterwards, the predators seize pupae from the attacked colony and transport them, including the female and the male pupae, into the *Harpagoxenus* nest.

New *Harpagoxenus* colonies result by means of inseminated females creeping into *Leptothorax* colonies, and once there, suddenly proceeding to cut off the legs and feelers of the inhabitants with their powerful toothless mandibles, until they are eventually lying with their extremities removed (Fig. 9.1). After this, the female takes possession

Fig. 9.1   Colony founding flight of *Harpagoxenus sublaevis*. The parasite female (H) can be seen, and a worker (L) with its feelers and legs cut off, as well as the removed extremities lying around (Photo — A. Buschinger).

of the brood and waits for her future slaves to hatch before she herself
begins to lay eggs.

A close relation of *Harpagoxenus*, *Harpagoxenus americanus*, founds
its colonies in a very similar manner to the species indigenous to parts
of Europe (Sturtevant, 1927; Creighton, 1929; Wesson, 1939). The
females of *Harpagoxenus americanus*, however, all have wings and their
helper ant species is either *Leptothorax curvispinosus* or *Leptothorax
longispinosus*. In the predatory expeditions of *americanus* observed by
Wesson in the natural environment, larger larvae, and even fully
developed workers were taken as well as worker pupae, but not the
pupae of sexual animals.

The American species *Leptothorax duloticus*, which is also dulotic,
has specialized even less in the predation of pupae than *Harpagoxenus
americanus*. This species goes out to steal pupae earlier in the year than
Harpagoxenus americanus, as a result of which it takes mainly eggs and
larvae from the nests of its helper ant species (Wesson, 1940). The
predatory expeditions of these social parasites are organized so that the
scouts lay a trail to their own nest, thus procuring reinforcements for
the attack.

According to more recent observations by Hölldobler (1976b), there
also exists among ants intraspecific dulosis, that is, where slaves are
stolen from alien colonies of the same species. A unique example of this
was found by Hölldobler among the honey ant species *Myrmecocystus
mimicus* in south-west USA. Alien colonies are attacked by this species
when they are smaller than their opponents of the same species. The
workers of the larger colony overrun the alien nest, there killing the
queen and dragging all the larvae, pupae, freshly hatched workers and
honey pots back with them into their own nest. This plundering of an
alien nest may extend over a number of days.

9.2 Social parasitism without dulosis

The permanent social parasites without dulosis mostly share one thing
in common, that the workers of these species, where they exist at all or
have ever been present in the species, present a pretty feeble existence.
Their number can be reduced without incurring any loss to the colony,
they may even die out completely without the colony being damaged in
any way.

These useless workers, which are fed by the helper ants and carried away by them in times of danger, are found in the genus *Epimyrma*, which is a parasite of *Leptothorax* species. Of the eleven Epimyrma species known so far, five have a worker caste; the remaining species either do not have workers, or they have not yet been discovered (Kutter, 1973b). The females of the parasites are inseminated inside the mother nest; here they rid themselves of their wings, leave their home colony and seek out a helper ant colony.

The females of *Epimyrma gösswaldi*, a parasite ant of the Main area, look for *Leptothorax unifasciatus* colonies and appease the hostile workers by stroking them soothingly with their feelers. In this way they obtain entry into the nest and once there seek out the queen and kill her from behind with their sharp mandibles (Fig. 9.2). After this they

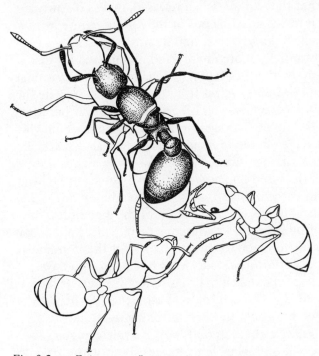

Fig. 9.2   *Epimyrma gösswaldi* female killing a host ant queen *(Leptothorax unifasciatus)* and two *unifasciatus* workers (after Linsenmaier, 1972, adapted).

are adopted by the *Leptothorax* colony and begin to lay their eggs (Gösswald, 1930, 1933, 1934).

The adoption of *Epimyrma stumperi,* found in Switzerland, was investigated by Kutter (1951), and described (1969) in the following manner: 'An *Epimyrma stumperi* female... is able to obtain its goal in a variety of ways. She first creeps forward with the utmost care, then suddenly plunges towards the danger, plays dead temporarily by stopping short, then suddenly runs away. This is repeated until the hostilities die down. Frequently, however, *Epimyrma* proceeds in a more sophisticated fashion. When an opportunity presents itself she will suddenly seize a *Leptothorax* worker by the neck, mount its back and immediately begin rubbing the captured ant with the special bristles on her forelegs, whereupon she then rubs herself with the same bristles all over with pedantic thoroughness. In this way, she undoubtedly achieves a direct transfer of the *Leptothorax* scent from its bristles on to her own body. The female thus perfumes herself, simultaneously camouflaging herself with the scent and also favouring the captured ant with her own *Epimyrma* scent, so that the worker, when released, makes its own unwilling contribution to the levelling out of the *Epimyrma-Leptothorax* scent difference. In fact, *Epimyrma* can find its way into the *Leptothorax* nest substantially more rapidly and unhindered by this procedure. Without any further hesitation, it then looks for the primary *Leptothorax* queen and promptly seizes it by the neck. The female then either rolls or throws the queen over on her back, stands over her again and bites her hard in the gullet. The significantly larger and seemingly stronger *Leptothorax* queen hardly defends herself against this attack. She seems not to realize the significance of the enemy, which is now trying to stab through the intersegmental membrane between the head and chest of her victim with her sharp, sabre-like mandibles. Since the excretory duct of a gland opens close to the tip of the mandibles, she has to try to spray poison into the wound immediately after piercing it. This nevertheless appears to involve some difficulty for the murderer: she may hold down the *Leptothorax* queen for hours, or even days. Eventually she will release the queen in order to recover herself. She will want to drink water and clean herself, only to seize the *Leptothorax* afresh and throw her on her back. In the intervening time, the *Leptothorax* may well have righted herself, but as a rule a queen maltreated in this way will only be able to stagger about with difficulty. Only when she is finally lying still will the *Epimyrma* release her and turn her attention to the next *Leptothorax* queen. The behaviour of the workers during all this was, and continues to be, variable: it can change from complete indifference to great interest. They hardly ever actively turn against the *Epimyrma,* but may persistently snarl themselves up around the unequal pair of combatants in a hopeless immobile tangle, rather like fascinated spectators. Since the host species *Leptothorax tuberum* is often polygynous, several queens may live peacefully together in the same colony; since, however, *Epimyrma* appears to be strongly monogynous, they take on a murderous attitude towards the females in the *Leptothorax* nest and against all host queens. Winged, that is non-inseminated queens, are never attacked, even if they have lost their wings. The elimination of a number of *Leptothorax* queens can, as will be readily seen, overtax the strength of an *Epimyrma stumperi* female to the extent that she will become prematurely exhausted. One may ask with some justification, therefore, why the *stumperi* do in fact kill all the host queens of a colony one after the other with such self-destructive zeal. The *stumperi* behaviour seems all the more tragic in the light of the fact that the *tuberum* queens are completely indifferent to a *stumperi* female. These females have nothing to fear from the *tuberum* queens in the way of competition and could easily dispense with the protracted and laborious murder of the queens, in other words content themselves with adoption by the *Leptothorax* workers.'

All the remaining *Epimyrma* species are parasites of monogynous *Leptothorax* species, and therefore have less difficulty in being adopted. Among *Epimyrma gösswaldi* and *Epimyrma stumperi,* which kill their host queens, larger workers of the helper species begin with the egg-laying and produce workers as offspring, according to the observations of Gösswald. If this is the case, then the queenless helper ants are themselves contributing to the maintenance of their stock for the benefit of the parasites. These observations by Gösswald, however, were to be checked again. Gösswald has in fact observed the development of *Leptothorax* workers in their *Epimyrma* nests over a period of months, including a winter rest period. He did not know at that time, however, that some *Leptothorax* larvae go through the winter period twice, after which they may still develop into workers. The possibility thus exists, that the *Leptothorax* workers produced in the *Epimyrma* nests originate from the brood that was taken over during the founding of the colony (Buschinger, pers. inf.). *Epimryma ravouxi,* on the other hand, has apparently dispensed with the killing of the helper ant queen. He states that here the legitimate queen of the helper ant colony sees to the continuance of the helper ants. It may also be the case, however, that the 'host queen' in the single observation presented here was not inseminated, but this was not investigated at the time.

The two socially parasitic species *Anergates atratulus* and *Teleutomyrmex schneideri* are both completely without a worker caste of their own: they may thus be regarded as the epitome of social parasitism. Compared with *Epimyrma*, they display further degenerations apart from the loss of their worker caste, which could only have been incurred as a result of their parasitic lifestyle. *Teleutomyrmex* has evidently reached the ultimate level of parasitic degeneration, as has already been stated in the description of the species by Kutter, who discovered it in 1950.

Anergates atratulus has been a known species for over 100 years. It lives, like *Teleutomyrmex,* among the grass ant species, *Tetramorium caespitum. Anergates* females resemble the free living species in their external appearance; only their stinging apparatus is completely reformed. The males of the species, on the other hand, are hardly recognizable as ants any more (Fig. 9.3). They are a dirty white in

Fig. 9.3   (a) male; (b) female of the workerless socially parasitic species *Anergates atratulus.* The female is already showing marked physogastry a week after insemination (Photo – A. Buschinger).

colour and are more resminiscent of pupae than fully developed ants. The rear section of their bodies is bent towards the front and they have lost their wings during the course of their evolution. Meyer (1955), who investigated *Anergates* from the histological standpoint, found extensive reformations in the internal structure of the males also, comprising whole sections of the digestive tract, a number of glands and, in particular, the skeletal musculature and the brain. Such feeble males are unable to leave their nest and inseminate females from other colonies. In the home nest, however, they are astonishingly active in the insemination of their female siblings. A proportion of the inseminated *Anergates* females remain behind in the nest, but the majority fly away and try to be adopted by alien *Tetramorium* colonies.

The chances for the *Anergates* females of being adopted by a *Tetramorium* colony are not good. The adoption very probably succeeds only in those cases where an *Anergates* female comes across a queenless *Tetramorium* colony, and this cannot happen very often. In *Tetramorium* colonies which do have a queen, and in which a strict monogyny prevails, the *Anergates* female would have to remove the legitimate queen, of which she is probably incapable. According to Kutter (1969), it was in fact observed on one occasion that *Tetramorium* workers killed their own queen in favour of an *Anergates* female (Crawley and Donisthorpe, 1912), but this would seem to be a rare exception to the rule. In the normal way of things, *Anergates* probably succeeds in being adopted only in helper ant colonies which are without a queen. This is supported by the observation of Gösswald, who was able to increase the presence of *Anergates* colonies in an area with many *Tetramorium* colonies by removing queens from grass ant colonies.

The meeting of the *Anergates* females with their helper ants and their acceptance into the colony is described by Gösswald (1954) in the following manner: 'When the *Anergates* female meets a *Tetramorium* worker, she is felt with its feelers outside the nest, and is then attacked. At this, the *Anergates* female lies down on her side, presses her legs to her body and rolls into a ball, remaining thus motionless on her back. Only her feelers move around in all directions; the jaws are wide apart. The *Anergates* females rolled up in this manner, in contrast to the moving females, arouse no hostility, in fact they are thoroughly touched and licked. If during this a feeler of the grass ant worker should happen to come between the jaws of the *Anergates* female, then the jaws will snap shut and the feeler will be held fast by its thicker end. The worker is thus temporarily paralysed. After it has recovered itself a little, it goes back into the nest to the brood with its feeler still clenched in the jaw of the *Anergates* female accompanying it, whose main goal would seem to be the brood. As soon as an attack ensues from the side of the *Tetramorium* workers, she rolls into a ball again. She goes around the nest in this manner for a number of days.' The presence or absence of a *Tetramorium* queen is thus the decisive factor governing the adoption of *Anergates* into a grass ant colony.

In those cases where adoption is successful, a race against time begins for the *Anergates* female. One can see that the helper ants will die out without the *Tetramorium* queen, and this time remaining is at the disposal of the *Anergates* female for her own reproduction. She uses this time to produce enormous quantities of eggs, her body swelling to extreme proportions within a few weeks (Fig. 9.3).

Teleutomyrmex schneideri is — judging from the finds so far — an extremely high alpine species. Kutter was the first to discover it in Saas-Fee at a height of about 2000 m and Collingwood, rediscovering it later (1955) at Briançon (Hautes Alpes) found it there at about the same height. This socially parasitic species is comparable with *Anergates* in a number of ways. They share in common a complete absence of a worker caste, cohabitation with the grass ant *Tetramorium caespitum* and the feeble structure of their pale males which are unable to fly. The *Teleutomyrmex* males do in fact still have wings, but they are reduced to such an extent that the animals are no longer able to fly and are similarly unable to leave their home nest. Insemination takes place in the mother nest, but not necessarily between siblings, since *Teleutomyrmex* is a polygynous species and the females may originate from different nests. In the degeneration of the internal organs also the *Teleutomyrmex* males ressemble the *Anergates* males, whilst the *Teleutomyrmex* females are more markedly degenerated in this respect than the *Anergates* females. The mouth parts in particular are reformed, as well as the musculature around them, the cleaning apparatus of the legs, a number of glands, especially the labial and metathoracal glands, the stinging apparatus and some parts of the brain, such as the corpora pendunculata (Brun, 1952; Gösswald, 1953).

On the other hand, however, *Teleutomyrmex schneideri* displays a number of physical features typical of adaptation to their particular parasitic lifestyle. *Teleutomyrmex* lives in the grass ant colonies as an ectoparasite on the backs of its hosts (Fig. 9.4). This adherence to their hosts by the parasites is facilitated by their legs, which are able to spread particularly widely round the host, and which are equipped with appressoria. In addition, the gaster is broad and plate-shaped, as well as being well covered with hairs on the underside. Inseminated *Teleutomyrmex* females have a stronger 'clasping drive' than non-inseminated; they favour the rougher surface of the thorax of the host queen to the smooth rear segment, but they also ride on the backs of workers and may even occasionally be seen still mounted on the backs of dead animals. It is possible that this holding fast on the backs of their hosts plays a part in their dispersal, by inseminated females allowing themselves to be taken on the nuptial flight by *Tetramorium* females. On the other hand, the parasite females are able to fly quite well themselves, so it is equally likely that they start out alone and then

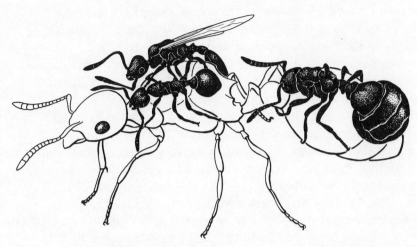

Fig. 9.4   Females of the workerless, extremely socially parasitic ant species *Teleutomyrmex schneideri,* on the back of a host queen *(Tetramorium caespitum)* (after Linsenmaier, 1972, adapted).

sink to the ground and allow themselves to be carried into the nests of *Tetramorium* workers which happen by.

The adoption of *Teleutomyrmex* females into *Tetramorium* colonies seems to ensue without any great dangers being involved. The parasite females are accepted there willingly after only a small amount of hesitation; glandular secretions from the parasites play a very important part in this ready acceptance. These glands have been found under glandular hairs and chitin pores on the thorax, petiolus and the first segment of the gaster. The parasites are licked very thoroughly and for a long time at these points. The *Teleutomyrmex* females are also able to eliminate any initial hostilities on the part of the grass ants by means of pacifying touching with the feelers, and by mounting workers, from whom they take on the scent of the colony.

Another group of workerless social parasites shows no apparent signs of degeneration. In external appearance its members are similar in every respect to independently living species; even as regards their internal organization, the degenerations were not looked upon as involving any serious losses for a long time. As a rule, these parasites are very closely related to their hosts and generally belong to the same genus. Disregarding the fact that these parasitic ants, grouped together by Kutter (1969) as 'Collective II', do not produce any workers, then they differ from their hosts only in very insignificant features. The typical characteristics for these parasites are simply a ventrally laid spine process or acantha at the postpetiolus ('parasitic spine'), a less sturdy formation of the body surface and a neater body structure.

A typical example of this group of parasites is *Leptothorax kutteri,* discovered by Buschinger (1965) in the Nürnberg area and also later found in Switzerland. *Leptothorax kutteri* lives with *Leptothorax acervorum* and only differs very slightly from its host in external characteristics. Despite the slightness of this difference, however, there is no doubt that *Leptothorax kutteri* is a distinct species, since it mates only with its own kind and never with its host species. The inseminated females are accepted without great difficulty into colonies of the host species, although they may meet with a small amount of hostility. *Leptothorax* females themselves when entering polygynous colonies will have to reckon with similar hostilities. The *kutteri* females, like the *acervorum* females, leave the queens of their helper ants unmolested. They fit into the group of already existing queens and allow their brood to be reared by their host workers.

The variety and extensiveness of the degenerations among *Anergates atratulus* and *Teleutomyrmex schneideri* allow one to conclude that there has been a long period of evolution towards parasitism, at the beginning of which were probably independent species. The involution processes concerned not only the various characteristics of the sexual animals, but led also to a reduction of the worker caste which is already far advanced among *Epimyrma*, and which has reached the final stage among the workerless *Anergates* and *Teleutomyrmex* species.

The course of development probably proceeded differently among the non-degenerated social parasites of the *Leptothorax kutteri* species; the indications suggest that it must have split from its host species after a relatively short period. This splitting of the parasite from the host species can, according to Buschinger (1965), be envisaged for *Leptothorax kutteri* in the following way. The ability of social insects to produce workers is laid down in their heredity and may be lost in a loss mutation. Such mutated females, no longer able to produce workers, are naturally no longer able to found colonies independently. They are only able to survive if they — like *Leptothorax acervorum* — belong to polygynous species and may be adopted by colonies of the same species. If genetic isolation in relation to the original species is then added to the loss mutation, then in principle a social parasite has emerged which in its external appearance and lifestyle would correspond to *Leptothorax kutteri.*

Arnoldi (1930) had already discussed an evolutionary path to social parasitism in principle similar to this in connection with the species *Symbiomyrma karavajevi* first described by him, a species which was later placed in the *Sifolinia* genus (Yarrow, 1968; Kutter, 1973a).

Doronomyrmex pacis is another non-degenerated social parasite. The first specimen of *Doronomyrmex* was discovered by Kutter (1945), on the day when war officially ended in 1945, at the peak of the Eggerhorn in Switzerland. On this occasion, a solitary female of this new species was walking over a stone, probably in search of a host colony. In consideration of the special day of its discovery, Kutter named the newly discovered species *Doronomyrmex pacis* (Gift of Peace). The next find of this 'peace ant' only succeeded, in spite of painstaking search in the area of the first find, as much as four years later, close to Saas-Fee (Kutter, 1950). This find provided the first males of the species and in addition insights into the behaviour of *Doronomyrmex*, which fell into a mould similar to *Leptothorax kutteri*. *Doronomyrmex* also allows the queens of its host species to live and does not hinder them from producing workers and sexual animals. *Doronomyrmex* — like *Leptothorax kutteri* — probably also evolved from its host species. The indication of such an origin for both species was interpreted by Kutter from a find made by him in the Saas-Fee area. There Kutter found a *Leptothorax acervorum* colony containing eight *acervorum* females which showed a clear ventral spine process at the postpetiolus, the parasitic spine in other words, which *Doronomyrmex, Leptothorax kutteri* and other social parasites have in common. These '*acervorum*' females, which, according to Kutter, represented three transitional mutations from *Leptothorax acervorum* to *Doronomyrmex*, were later interpreted by him (1967) to include the species *Leptothorax gösswaldi* (mutant I), *Leptothorax kutteri* (mutant II) and *Leptothorax buschingeri* (mutant III). Whilst *Leptothorax kutteri* and *Leptothorax gösswaldi* are definitely distinct species, *Leptothorax buschingeri*, of which we so far know only the males, may represent only an extreme variation of *Leptothorax acervorum*.

It is possible that the way is being prepared for social parasitism among *Myrmica rubra* and *Plagiolepsis pygmea*. In both species two different forms of sexual animals were found, differing substantially in size. Faber (1969) investigated these different sexual animals among *Myrmica rubra* in Nordrhein-Westphalia and found there not only the small female forms already known (microgynes), but also males of different sizes. These males could be divided into two main size groups, which were linked by intermediary forms. In the area investigated, the microgynes did not go to all places in equal numbers, but tended to go to specific points. All colonies with females of normal size only contained normal sexual forms as offspring. Faber (1969) also found microgynes in association with normal females. In these colonies, the sexual forms produced consisted exclusively of microgynes and smaller males. It may be that the facultative normal females of these mixed colonies are prevented from attaining their normal size. Faber concluded from his discoveries that among *Myrmica rubra* two separate reproductive cycles have already formed, in which the microgynes are on the way to becoming social parasites which will be the parasites of its original species.

Among *Myrmica ruginodis,* small forms of males and females were found which tend to mate among themselves (Brian and Brian, 1955). Here also, a path of evolution to social parasitism could be forming, as Elmes (1976) has also discussed in connection with deviant forms in *Myrmica sabuleti* populations. Elmes (1973b, 1976), however, does not take the microgynes of *Myrmica rubra* to be the precursors of future parasites. Corresponding to the conclusions of Hauschteck (1965) from *Myrmica sulcinodis* and Crozier (1970) from numerous ant species, Elmes assumes that the larger female forms in *Myrmica rubra,* with their 48 chromosomes, are the result of polyploidization (multiplication of the chromosome set). The partial correlation which exists between the size of individual ants and the number of their chromosome sets is taken by Elmes as an indication that the microgynes represent an earlier stage of evolution in which there was a smaller number of chromosome sets. The microgynes of *Myrmica rubra* could thus have resulted spontaneously by means of a reduction in the number of chromosome sets, and therefore represent not a development into future parasites, but in fact earlier stages of evolution.

Microgynes of the species *Plagiolepsis pygmaea* were found by Reichensperger (1911) in two *pygmaea* colonies near the Moselle. He was able to observe with these abnormal females that they laid eggs in observation nests from which normal workers would develop. Even then Reichensperger indicated that the possibility should not be ruled out of the microgynes described representing a 'new formation and branching-off from the *pygmaea* species'.

Kutter (1969) advances further examples of workerless, non-degenerated social parasites. The majority of them have still been little investigated, for example, *Myrmica myrmecoxena,* discovered by

Bugnion in 1874 in a *Myrmica lobicornis* nest in Anzeindaz or
Gryon-sur-Bex at a height of 1900 m, but which was never discovered
again. Equally little is known of the life habits of *Myrmica
myrmecophila,* of which Wasmann found a female in a *Myrmica
sulcinodis* colony on the Arlbergpasshöhe. *Camponotus universitatis,*
another species which probably belongs in this group, was rediscovered
twice after being first described by Forel in the year 1890; all we know
of its biology is that this species lives in *Camponotus aethiops* colonies.

Leptothorax gösswaldi is known to be another species which, like
Doronomyrmex and *Leptothorax kutteri,* lives in the colonies of
Leptothorax acervorum (Kutter, 1967). Also belonging in this group of
social parasites are what have been designated by Faber (1969), the
'begging ants' — in contrast with the assassin ants, which eliminate their
host queens — a number of social parasites of *Plagiolepis,* in particular
Plagiolepis pygmaea and *Plagiolepis vindebonensis.* These little,
dwarfish ants are inhabitants of xerothermic regions in central and
above all, southern Europe.

The first to describe the social parasites of *Plagiolepis* was Stärcke
(1936), in describing the females of *Plagiolepis xene,* which are
evidently able to find shelter in the colonies of *Plagiolepis pygmaea*
and *Plagiolepis vindebonensis* if they so choose. Kutter discovered the
males of this species in 1952, which are wingless and similar to the
females in their external appearance to such an extent that it is difficult
to distinguish between the two sexes if the initially winged females have
shed their wings. Insemination among these parasites takes place inside
the nest among siblings. The inseminated *xene* females may be adopted
by the alien *pygmaea* or *vindebonensis* colonies. According to
laboratory observations made up to now, adoption is successful only in
the face of great difficulty. It is worth noting here that the *xene* females
seek to place themselves under the protection of the host queens, which
are far more amenable towards them than the workers. It is thus less
dangerous for the parasite females if they attach themselves to groups of
colony-founding host females. According to Passera (1964), the
parasites probably spread themselves out by following groups of host
ants outside their nests and thus being taken over from the mother
colony into branch colonies.

Plagiolepis grassei was discovered by Le Masne (1956b,c) in the
eastern Pyrenees to be a parasite of the species *Plagiolepis pygmaea.*
Systematically, this species seems to stand between *Plagiolepis
pygmaea* and *Plagiolepis xene,* but it does not quite fit into our group

of social parasites, inasmuch as it produces a number, albeit a small number, of workers. On the other hand, the *Plagiolepsis* parasites discovered by Karawajew (1931) in Russia, by the Caspian Sea, and by Faber (1969) in upper Wachau (lower Austria) were without workers. The parasites discovered by the two authors belonged to different species. Karawajew named the species which he discovered *Plagiolepsis regis,* and described it as an independent species. He overlooked the fact that his find actually consisted of two different species put together, the workers of the host colony and the workerless parasites, among which he found females and small, wingless males. The social parasites discovered by Faber differed so markedly from their host species *Plagiolepsis pygmaea* and *Plagiolepsis vindebonensis*, as well as *Plagiolepsis xene* when investigated more closely, that Faber placed them into a new genus and named them *Aporomyrmex ampeloni.* He also placed the parasites discovered by Karawajew in this genus, so that they were now called *Aporomyrmex regis.* Among *Aporomyrmex ampeloni,* Faber discovered a rare case where four different forms capable of reproduction occur, winged males, winged females, and also wingless forms of both the sexes. Distinguishing between the sexes of each of the corresponding forms is no easy matter. Externally, sexual dimorphism extends to insignificant differences in size and the shape of the chewing edge of the mandible. In contrast with the females, the two male forms differ from each other in terms of numbers: Faber found only a single pterane (winged male) and assumed, in view of the large numbers of aptanes (wingless males) he found, that the rare pteranes represent an atavism, i.e. a regression to a male form that was previously more abundant, if not exclusive. Among *Aporomyrmex* also, insemination takes place in the mother nest; the most numerous and extremely active male form here makes no distinction between winged and wingless females. They also try their luck with workers of the host species, and even with other males of their own species. The greater number of males to females and the marked readiness of the males to mate is sufficient to ensure that every member of the parasite species is inseminated. After mating, the greater proportion of the females leave the nest. The adoption of these females by an alien colony of helper ants would seem — according to laboratory observations of Faber — to occur in increasing proportion to the decrease of their helper ants. By contact with a few workers of the host species and their waste material the parasitic females may acquire a protective odour and thus gain unharmed entry to the nest.

A particularly large number of social parasites of the 'begging ant' type have emerged from the *Myrmica* genus. Many of these parasites have only been discovered in recent years and many questions regarding their lifestyle remain open. Apart from the *Myrmica* parasites already named, other species belonging to this group are *Myrmica lampra*, a parasite of *Myrmica kuschei,* discovered by Francoeur (1968) in Quebec, *Myrmica faniensis* (van Boven, 1970), *Sommimyrma symbiotica* (Menozzi, 1924), of which only a single female was found in the Appenines, south of Modena, and a few *Sifolinia* species. *Sifolinia kabylica* was discovered by von Cagniant (1970) in *Myrmica aloba* colonies in Algeria, together with reproductive females of the host species. Of *Sifolinia lemasnei* one female only was originally described as *Myrmica lemasnei* (Bernard, 1968), which had been discovered by Le Masne in 1950 in a *Myrmica sabuleti* colony in the Pyrenees. This first find went missing, but Le Masne later found further females and also males of this parasite species, which after Kutter (1973a) was described as *Sifolinia lemasnei.* Of *Sifolinia winterae*, Winter (1973) first discovered a wingless female as the first specimen, in Switzerland at at height of 1400 m, the female being similar in its external appearance to a *Myrmica microgyne.* Shortly after this discovery, Buschinger discovered further females, together with host workers of this new species and was thus able to show that it was a parasite of *Myrmica rugulosa.* Buschinger kept these parasites with their hosts for over a year in the laboratory, but was unable to obtain any offspring from the parasites and only males of the host species hatched from worker eggs.

The list of social parasites could be expanded with many more species, especially if one were to include species from North and South America, such as *Manica parasitica,* which is a parasite of *Manica, Pogonomyrmex anergismus,* which is a parasite of *Pogonomyrmex, Sympheidole elecebra* similarly a parasite of *Pheidole, Aphaenogaster tennesseensis* of *Aphaenogaster* and *Pseudoatta argentina* of the leaf-cutter ant *Acromyrmex.* Social parasites of the begging ant type are also known in Australia: the first specimen was discovered by Brown (1955). He found the new parasitic species *Strumigenys xenos* in a young *Strumigenys perplexa* colony, a species belonging to the Dacetini, which is found distributed in the moist southern parts of Australia, Tasmania, northern New Zealand and a number of neighbouring islands *Strumigenys xenos* was found in nests with reproductive females of the host species and seems to be accepted by the majority of the polygynous colonies of its host species without undue difficulties. The diet of these workerless parasites has remained unknown for a long time, as well as their reproduction and dispersal.

Another parasite from the Dacetini group, a parasite of *Strumigenys loriae,* was found by Wilson (Wilson and Brown, 1956) in New Guinea. This parasite belongs to the *Kyidris* genus, until then known to exist only in Japan and Formosa. The *Kyidris* females leave the host queens unharmed and thus belong to the begging ant group. They fall outside the range of these social parasites, however, inasmuch as they produce workers, albeit not particularly capable ones. They are practically incapable of hunting prey or participating in building operations and only manage to take the brood to safety very slowly if the nest is opened. On the other hand, these workers, which are nevertheless very numerous, show only very slight signs of morphological degeneration. This leads us to conclude that in the case of the *Kyidris* species we are dealing with a relatively young parasite from the phylogenic point of view.

It is worth noting that in Australia, even among the primitive *Myrmeciines,* a social parasite was found. The find goes back to Douglas (Douglas and Brown, 1959), who found the parasitic species *Myrmecia inquilina* in the region of Badjarning Rocks in a *Myrmecia vindex* nest. The *Myrmecia inquilina* females kept themselves close to the brood and the legitimate host queen. *Myrmecia inquilina* was later rediscovered in the same region (Haskins and Haskins, 1964). This new discovery showed that the parasites were accepted without difficulty by their host colonies. The host queens continue to produce sexual animals and workers in the presence of the parasites, although in reduced numbers. We still do not know to what extent the presence of the parasites alone is responsible for the reduction in the numbers of offspring. According to Haskins and Haskins (1964), this reduction is probably the result of an abuse of the hospitality of their hosts by the parasites, by consuming the eggs produced by the host queens.

10 Xenobiosis, lestobiosis, parabiosis and kleptobiosis

Xenobiosis describes the relationship of guest ants with their hosts. Guest ants are not social parasites, although, like them, they also live with other ant species and would not be capable of living without them. They are not counted among the social parasites because they have preserved a certain independence for themselves. In the nests of their hosts, they maintain their own 'household' and rear their brood without help from outside. They receive only board and lodging from their hosts. To see how the cohabitation of these obligatory guests and their hosts looks, we shall present the example of *Formicoxenus nitidulus* in some detail.

Formicoxenus is the only known guest ant species in Europe. The colonies of this species, which with less than 100 individuals are relatively small, are in adapted hollows in the top of wood ant nests. The particular noticeable feature of these 'shiny guest ants' is that the castes are not sharply distinguished from one another. One can find among them continuous transitions between females and workers, where the intermediary form is represented by fully developed females which may also be inseminated; only one female is ever fertile in one colony at any one time, however (Buschinger, 1976a). The males of *Formicoxenus* are wingless and resemble workers in their external appearance. It is nevertheless worth noting that the 'parasitic spine' is present among these guest ants, a feature otherwise typical only for specifically socially parasitic species. This characteristic of *Formicoxenus* probably has no significance for its lifestyle, however, since other guest ants found in Africa lack this additional feature.

The life habits of *Formicoxenus* were thoroughly investigated by Stumper (1918, 1921b) and by Stäger (1919, 1923, 1925). They noticed that the guest ants remain almost exclusively inside their host nests and are only very seldom seen in the open. This led them to suppose that *Formicoxenus* not only obtains a place in which to nest from the wood ants, but that they also receive food from their hosts. Stumper assumed first of all that the diet of the host ants consisted of

organic and inorganic material picked up by rain water as it seeped
through the ant hill. This hypothesis could not be proved, however.
Instead, Stäger observed, the guests allow themselves to be served by
their hosts by participating in the social food flow. This occurs either by
begging and allowing themselves to be fed by them, or else by taking
the opportunity to be fed, where two of their hosts are mutually
feeding: they clamber quickly on to the head of one of the participating
animals, thus obtaining a portion of the food being handed on. The
wood ants tolerate their unbidden guests, by whom they seem to be
neither particularly harmed nor disturbed. Only occasionally do they
attack their guests, which then wait motionless and cowering until the
attack is over. Only very seldom will the guests pierce them with their
dangerous sting.

Other guest ants, whose lifestyle is very similar to the European
Formicoxenus, are known in North America. *Leptothorax emersoni,* for
example, a species discovered by Wheeler (1903, 1904a, 1919) and
investigated by him, lives with *Myrmica brevinodis.* The ability of the
guest ants to join in the social flow of food was first discovered with
this species. They climb on to the backs of their hosts and trill their
heads with their feelers until a drop of liquid food appears. *Leptothorax
diversipilosus,* a species discovered by Smith (1939, 1956) and
investigated in more detail by Alpert and Akre and *Leptothorax
hirticornis* (Smith, 1939; Snelling, 1965) are two more guest ant species
which were found in America. Both species nest with *Formica
obscuripes.* They are similar to our guest ants as much in their lifestyle
as in the formation of intermediary forms between the females and the
workers. The males in these species also are wingless and very similar in
appearance to their workers. In contrast with *Formicoxenus,* however,
they possess no spine process at the postpetiolus. They were not able to
observe mating among the American guest ant species in the open;
Alpert and Akre (1973) nevertheless collected winged females as well as
males belonging to the same species in the direct vicinity of the host
nests. This led them to conclude that the mating of *Leptothorax
hirticornis* females with their wingless males takes place not only inside
the host nest, but also in the area surrounding it. The inseminated
females either seek out a new host nest or return to the nest from which
they came.

The relationship which maintains a number of ant species as 'thieving
ants' is described as lestobiosis. The only thieving ants indigenous to
Europe are species of *Solenopsis* such as *Solenopsis fugax,* whose

behaviour is known mainly from the work of Forel (1869), Wasmann (1891) and Hölldobler (1923, 1928, 1965). *Solenopsis fugax* lives in the warmer areas of central Europe, and is quite prolific in places. The tiny yellow *Solenopsis* workers bore fine passages into the nests of larger ant species, and there seize the brood and any other valuable food source. Afterwards they disappear back into their passages, where the larger ants cannot pursue them. The chemical strategy which they use at this time has already been described (*see above*). The ant species sought out by *Solenopsis* are, among others, *Formica fusca, Formica rubifarbis, Formica sanguinea, Polyergus rufescens, Lasius niger, Lasius alienus*, various *Myrmica* species, *Tetramorium caespitum* and *Tapinoma erraticum*. It is also possible that we may have to distinguish between two different types of *Solenopsis*, which, according to the investigations of Hölldobler (1965) show substantial deviations in behaviour. In contrast with earlier investigations of known *Solenopsis fugax*, which conducted their thieving expeditions, but otherwise left the colony from which they stole in peace, Hölldobler found one *Solenopsis* colony which conquered individual sections of a nest or even whole nests by first killing the adult animals of the colony and then eating the brood.

Other thieving ant species are known to exist in the *Carebara* genus distributed in Asia and Africa, and which in their thieving lifestyle overrun termites in particular. *Monomorium pharaonis*, which is today very widely distributed as a species of house ant, lived originally as a thieving ant species in its East Indian indigenous home.

Parabiosis describes a form of cohabitation of ants in which different ant species make use of a single trail. This extremely free form of cohabitation is very rare among ants, only being known to exist in a few species. Parabiosis is nevertheless parasitic in one known example (Kaudewitz, 1955); other examples of parabiosis have a more indifferent (Wilson, 1965), or only very slightly symbiotic character (Wheeler, 1921; Weber, 1943).

The concept of 'parabiosis' stems from the work of Forel (1898), who used it to describe the relationship between *Crematogaster linata parabiotica* and *Monacis debilis*. These two species live in the South American rain forests, usually nesting in separate nests in close proximity to each other which are connected by means of passages. Whilst each of the two species nevertheless remains alone inside the nest, they make use of the same trails in the search for food outside the nest. A similar form of cohabitation was found by Wheeler (1921) between

Crematogaster and *Camponotus femoratus.* Here also the two species use the same trails, displaying no hostility of any kind in their use; on the contrary, they greet each other with gentle touches of the feelers and even feed each other. Weber (1943) moreover discovered that the two species even defend their nests together.

In the parabiosis observed by Kaudewitz (1955) between *Camponotus lateralis* and *Crematogaster scutellaris, Camponotus* follows trails laid by *Crematogaster,* simultaneously carrying with it the excreta of plant-eating mammals. Any hostility shown by *Crematogaster* ants is not reciprocated by *Camponotus*: the *Camponotus* workers curl up into a ball during the attack and go into a 'waiting posture' until it is over.

The parabiosis observed by Wilson (1965) exists between the more prolific species *Azteca chartifax* and the rare species *Camponotus beebi,* both found in Trinidad. Here also, *Camponotus* uses the alien trail; the two species nevertheless meet each other relatively rarely on their commonly used trail, since *Camponotus* goes in search of food mainly in the daytime, and *Azteca* chiefly at night. Although *Camponotus* generally moves along alien trails, this species is also able to lay its own trails, which are then, however, used only by *Camponotus* and not by *Azteca.*

Kleptobiosis describes a form of behaviour in which ants returning to the nest with food are robbed. This form of behaviour is known to exist in a number of beetle and fly species, but is also known amongst ants, which lie in wait near the trails of alien species and then take the food away from the returning ants. The robbers are mostly concerned with animal food, but occasionally also vegetable matter (Wroughton, 1892; Forel, 1901; Wheeler, 1910, 1936; Abe, 1971; Maschwitz and Mühlenberg, 1973a).

The kleptobiotic species *Camponotus rufoglauca* was observed in Ceylon in detail by Maschwitz and Mühlenberg (1973a). It was shown here that mainly workers of *Camponotus sericeus*, from nests of between 4 and 6 m in distance were robbed. Up to three robber ants would stay almost stationary in front of entrances to *sericeus* nests, moving around slowly. As soon as a returning worker approached the nest, the faster-moving and more agile thieves would fling themselves at it, thrusting forwards with their open mandibles and thus touching the head of the ant being attacked, and retreat immediately. The ant being attacked would never defend itself; it would either remain motionless or roll itself into a ball, thus allowing its captured prey to fall. This would then be quickly seized by the attackers and carried back into their own nest. *Camponotus sericeus* workers, which carry no such booty, were attacked in similar fashion. This indicates that the robber ants are not able to distinguish between laden and unladen workers. It could even happen that robber ants would be attacked in this way by their own kind as they were lying in wait near alien colonies. The *rufoglauca* workers also attacked expeditions of army ants of the species *Aenictus* as well as *Camponotus sericeus*. They would lie in wait beside the strongly marked trail and fall upon the small *Aenictus* ants, which would not defend themselves and would allow the attackers to take their booty.

The *rufoglauca* workers which lie in wait in front of alien nests seem to be quite specialized in their thieving lifestyle. This is shown by observation of one robber, which fell upon an artificially arranged feeding place of dead termites which had been taken by *sericeus* workers. It did not then begin to seize the termites lying around, but remained there to fall upon only the *sericeus* workers and take their booty from them. On the other hand, *Camponotus rufoglauca* colonies are quite able to feed themselves without help from outside and carry back animal food and honeydew independently (Maschwitz and Mühlenberg, 1973a).

11 The guests of ants

There is a group of species among the social parasites which has such a mastery of the 'language' of its hosts, that they are not treated like strangers, but rather as members of their own colonies. Not only ants have acquired this capability during the course of their evolution: there are other insects and arthropods also which have 'learned' to communicate with ants and to benefit from so doing. The more communication signals of their hosts which these Myrmecophiles or 'ant guests' are able to learn and master, and the better able they are to imitate these signals, the closer is their relationship with the ants and the more advantage they derive from their hosts.

The shiny beetle *Amphotis marginatus* knows essentially two of its hosts' signals and is able to roughly imitate one of these. It recognizes the trails of *Lasius fuliginosus* and waits beside these trails for ants returning to the nest. As soon as an ant with a full crop follows the trail, the beetle rushes up to it and begs for food (Fig. 11.1). It touches

Fig. 11.1 Shiny beetle *(Amphotis marginatus)* being fed by a *Lasius fuliginosus* worker, having begged on the 'road' (after Hölldobler, 1970b, adapted).

the head of the ant with its feelers and pushes its head against its mouthparts. These signals are usually sufficient to induce the ant to whom the beetle is begging, to regurgitate food from its crop, especially since ants with full crops will give food quite readily. More often, however, the worker will recognize that the begging beetle is an impostor and will begin to attack it. The beetle, however, which in the meantime has accomplished its goal, on being attacked withdraws underneath its strong back shield, presses the broad edges of the shield flat against the ground and then locks its feet into the ground with special bristles. The ant is thus unable to attack it and finally continues on its way (Hölldobler, 1968, 1973b).

Other Myrmecophiles which steal the ant brood as followers of army ants have already been mentioned. They are able to interpret the army ant trails and are protected, for their part, from attack by the ants by means of substances which have the effect of frightening them away. There are also harmless Myrmecophiles, however, such as beetles from the Deremini group of species which follow the trails of African driver ants, living apparently not on the ant brood, but only on the remains of prey and dead ants (Kistner, 1966, 1967). The Myrmecophile beetles of the *Paralimulodes* genus are also harmless. These tiny beetles ride on top of their hosts, placing themselves so skilfully in so doing that whole groups of them are arranged symmetrically on both sides of the head or rear segment of the ants, as the case may be (Wilson *et al.*, 1954). This phenomenon has also been described by Wheeler (1926) for myrmecophile mites and by Park (1933) for the *Limulodes* beetle genus. The diet of *Paralimulodes* and also of *Limulodes parki* consists of organic particles of waste shed from the cuticula of the ants. Whether or not the beetles eat ant broods in addition to this, it has been impossible to prove or disprove.

Among those beetles found to be followers of army ants, a few particular species from the family of Staphylinidae are known to resemble ants very closely in appearance (Fig. 11.2). However, this external similarity of the beetles to their hosts certainly does not help them to be accepted as equals by the ants: against this, for example is the fact that the army ants can hardly see at all or may be completely blind. It is far more likely that these beetles are protected from insect-eating birds by their similarity to ants. We take this to be the case from observations, according to which birds following armies of ants did not take the ants themselves, but other insects which were scared up by the ants.

The most well-known group apart from the ant-like beetles is that of the ant-like spiders, whose locomotion is also similar to that of ants to a

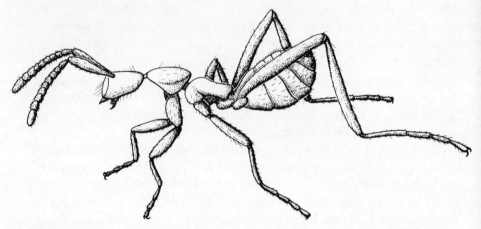

Fig. 11.2   Myrmeciton antennatum, an ant-like beetle belonging to the
Staphylinides and found following army ants (after Wilson, 1971 − from Seevers,
1965, adapted).

certain extent. In his book about mimicry, Wickler (1968) describes the
behaviour of these spiders in the following manner: 'There are spiders
which look deceptively like ants, and which also move around among
ants in shrubs. Spiders have eight legs, ants six; the ant spiders, however,
only walk on six of their legs and feel about with the foremost pair of
legs, as ants do with their antennae. A number of ant-spiders come from
the jumping spiders, which trick humans by suddenly disappearing from
a branch in a gigantic leap, at the same time drawing a thin safety
thread.

   One of the ant-like spiders belonging to the jumping spiders,
Synagales venator, a species found in central Europe, albeit rarely (Fig.
11.3), was investigated in more detail by Engelhardt (1970). *Synagales*
often uses its second pair of legs like feelers, keeping the first pair of
legs on the ground. In its behaviour, it resembles the ants, amongst
which it moves (*Lasius niger*) to an extraordinary degree. In order to
test whether in fact anyone is deceived by the similarity of the spiders
to the ants, Engelhardt conducted experiments with hand-reared
blue-tits.

   The blue-tits ate the pupae of ants in large quantities and evidently
with some relish. The ants which had hatched from the pupae, on the
other hand, were no longer accepted after the blue-tits had 'tried' them.
The spiders, however, which were offered by hand to the tits were eaten
by the birds immediately; but if the spiders were placed in a petri dish

so as to move freely among ants, then the blue-tits displayed only 'revulsion behaviour' by repeatedly whetting their beaks, but they did not eat the spiders. This would indicate that the spiders are protected amongst ants by their ant-like appearance and behaviour (Bates'

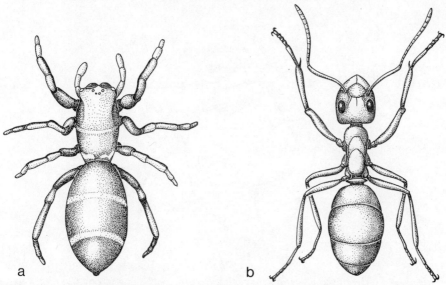

a   b

Fig. 11.3   The ant-like spider *Synagales venator* (a) and (b) *Lasius niger* worker (after Engelhardt, 1970, adapted).

mimicry). We do not know how effective this protection is against other predators, but even a limited protection against a few predators could mean a substantial selective advantage for the spiders.

The beetle species *Amorphocephalus coronatus* also occurs in central Europe, and lives in the nests of *Camponotus* species. Whilst forcing its way into the nests of its hosts it initially comes up against difficulty and is attacked by the ants. These attacks soon decrease in intensity, however, and the beetles are finally accepted by the ants and have the freedom of the nest. The species is fed by the ants (Fig. 11.4) and it has also been observed in this connection that the beetle itself hands on food to its hosts, something which is so far unique for ant guests (Torossian, 1965, 1966; Le Masne and Torossian, 1965).

There are in fact ant guests not only among the beetle species, but also in many other groups of arthropods. There are thus myrmecophile Crustaceans, Arachnids (*Pseudoscorpionada, Aranea, Acarina*), primitive insects (*Collembola, Thysanura*), cockroaches, Neuroptera, butterfly larvae and flies.

A closer analysis of the signals which make the guest-host relationship possible was carried out by Hölldobler using *Dinarda* as an example, as well as *Atemeles*.

Dinarda belongs to the Staphylinid group of beetles and lives among

Fig. 11.4   The ant guest *Amorphacephalus coronatus* being fed by *Camponotus* worker (after Le Masne and Torossian, 1965, adapted).

Formica species, mainly among the predatory ant species *Formica sanguinea*. Among the ants, the beetles move around in the nest, but do not penetrate the brood chambers. *Dinarda* lives in the peripheral portions of the nests, living mostly on the edible things which are to be found there, such as dead ants which have not yet been removed from the nest, and the remains of food brought into the nest. Occasionally the beetles also attempt to take part in the social flow of food within the ant colony itself, pilfering the liquid food passed on from one ant to another (Fig. 11.5). From time to time the beetles also beg from workers bringing food into the nest and induce them to hand over the food. In this instance, however, they are usually recognized as aliens by the ants and are attacked. The flat shape of the beetles nevertheless renders them very difficult to attack from the point of view of the ants, particularly when the beetles lie flat on the ground (Wasmann, 1889, 1934). According to Hölldobler, however, *Dinarda* possesses special glands in its rear segment, the secretions of which act to pacify the host ants. Whilst the ants are licking the pacifying secretion, the beetles are

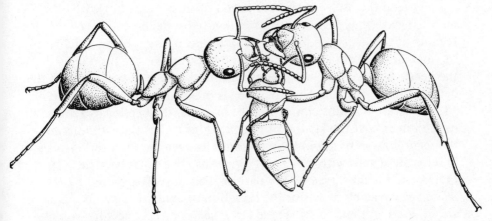

Fig. 11.5   The *Dinarda* beetle joining in the social food flow of *Formica* workers
(after Hölldobler, 1971c, adapted).

provided with the opportunity to escape without being attacked.

Far better adapted than *Dinarda* to their myrmecophile lifestyle are
the beetles of the *Lomechusa* group, which consists of the European
beetle genera *Lomechusa* and *Atemeles*, as well as the *Xenodusa* genus
found in North America and Mexico. The characteristic feature of this
group of ant guests from the Staphylinid family is that they have access
to the brood chambers of their host ants and that even their own larvae
are integrated with the ant colonies. Only the pupae of these species are
poorly adapted to life among the host ant species, as one can see from
their high mortality rate. Whereas in fact *Lomechusa* goes through its
entire development with one host species and later remains 'faithful' to
it, *Atemeles* and also *Xenodusa* display a regular changing of hosts.
Atemeles spends the summer, and thus the larval period, in *Formica*
nests and changes for the second half of the year including winter into
the nests of *Myrmica* species. *Xenodusa* on the other hand lives in
Camponotus nests in the winter and *Formica* nests in the summer
(Wasmann, 1910b).

The signals which make it possible for the beetles to be adopted by
various hosts, both as larvae and as fully grown beetles, were
investigated more close by Hölldobler (1967a,b, 1970a) using *Atemeles
pubicollis* as his example. The larvae of this species are found in the
field in particularly large numbers in *Formica polyctena* nests. The
larvae are kept by the hosts together with their own brood and are more
often licked and brought to safety sooner, when the nest is disturbed,

than the ants' own brood. The *Atemeles* larvae for their part initially happily feed themselves from the brood of their hosts, but with increasing age they are fed more frequently by the host ants. The feeding of the *Atemeles* larvae usually begins by their being touched by the maxillae and antennae of the workers caring for the brood. At this, the beetle larvae bend their bodies right back and seek contact with the heads of the ants by moving their heads in a pendulum motion (Fig. 11.6a). As soon as this contact has been made, the beetle larva presses its labium against the labium of the ant (Fig. 11.6b) thus stimulating the regurgitation of liquid food. During this feeding, the beetle larvae imitate the signals with which ant larvae also beg for food. The preference for the larvae on the part of their hosts is achieved by the *Atemeles* larvae by means of the secretion of epidermal glands which are situated on their backs. These larval secretions are acetone-soluble. If small pieces of filter paper are saturated with these extracts, then these filter papers will also be placed with the ant larvae. The beetle larvae are probably imitating the attractive substances of the ant larvae

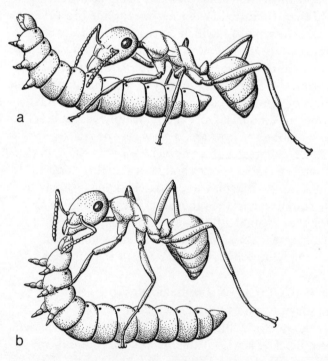

Fig. 11.6   Wood ant worker feeding an *Atemeles* larva. (a) the larva seeks contact with the ant by means of a pendulum-like motion and, when contact has been made, is then fed by the ant (b). The epidermal glands of the larva have been drawn in black (after Hölldobler, 1971c, adapted).

with these secretions from their epidermal glands. This is supported by the observation that *Atemeles pubicollis* larvae may be adopted by any ant species provided that their larvae may be interchanged with the ant larvae (Hölldobler, 1967a,b).

The nest change of the *Atemeles* beetles and the finding of new hosts takes place, according to Hölldobler (1969a,b, 1970a), in the following manner. The fully grown *Atemeles pubicollis* animals leave the *Formica* nest 6-9 days after hatching; during this time, the beetles seem to be in a 'lively mood' and display a good deal of movement activity. In this way they are led out of the forest biotope of *Formica* and into the field biotope of *Myrmica*. In addition, the beetles continue to orient themselves chemically, by reacting positively to the specific scent of the *Myrmica* species as well as recognizing and moving towards this scent even when it is mixed with the scent of other ant species. After the beetles have arrived at a nest of their winter host species, they do not enter immediately, but wait in front of the entrances to the nest until they are carried inside by the *Myrmica* workers.

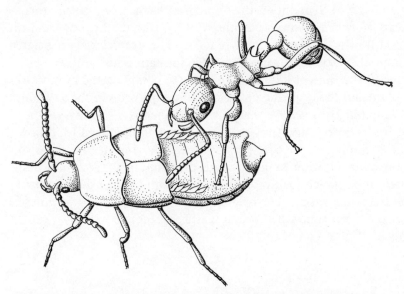

Fig. 11.7   Scene from the adoption of *Atemeles pubicollis* in a *Myrmica* colony. The beetle is licked by a Myrmica worker on the clusters of bristles which are connected with 'adoption glands' (after Hölldobler, 1970a, adapted).

The adoption of the beetles is usually successful, owing to the secretions of single glands which are situated to the upper side of the rear segment of the body (Fig. 11.7) and, in addition, 'pacifying glands' at the posterior end of the body. The *Atemeles* beetles are also equipped with a defence gland for emergencies, a gland also found among *Lomechusa*. Among *Lomechusa*, this gland contains mainly the

carbohydrate substance tridecane (Blum *et al.*, 1971), which is also
found as a defence substance in its host species, *Formica sanguinea*
(Bergström and Löfqvist, 1968). In the nests of *Myrmica* species, the
Atemeles beetles imitate the begging signals of their hosts so well that
they are fed as if they were of the same species (Hölldobler, 1970a).

According to Wasmann (1915), *Lomechusa* displays a migratory
phase in autumn (after hatching) and also in the spring (the time of
copulation), during which time the beetles leave the hosts' nest. Unlike
Atemeles, however, they do not change their host species, only the
colony, by being freshly adopted by another colony of the same species.
The migratory phases of *Lomechusa* take place at exactly the same time
as *Atemeles* changes its host species. According to Hölldobler (1967a,
1970a), the significance of the host change can be understood when one
compares the yearly cycles of *Formica* and *Myrmica* with the individual
development of *Atemeles*. In *Formica* colonies, the food supply is
greatest when the brood is being reared. This coincides with the
development of *Atemeles* in the nests of *Formica* species. From August
until the middle of September, the number of larvae in *Formica* nests
decreases — *Formica* colonies pass the winter without a brood — and the
social food flow is correspondingly reduced. The *Atemeles* beetles hatch
at this time, their spermatogenesis and egg formation not yet being
completed at the time of hatching. In order to complete the maturation
of the sperm and the eggs, the beetles require high-protein food, which
is to be found in *Myrmica* nests at this time, since *Myrmica* passes the
winter with its broods and thus also maintains an unreduced food flow
up to the beginning of the cold period. Since the *Atemeles* beetles also
eat ant brood in addition to liquid food, they find in the *Myrmica* nests
the protein food which facilitates the completion of their development
until the following spring. The beetles then return to the *Formica* nests,
in fact at the exact time when brood-rearing is recommencing
(Hölldobler, 1970a).

12 **Ant nests**

The ability of ants to maintain themselves in their various biotopes presupposes, amongst other things, that they have acquired the ability to use a wide variety of nesting places and adapt them to their own living requirements by means of special structures. The ingenuity of ants in their use of existing nesting possibilities and the skill applied by many species in the building of their nests are not very widely appreciated. Unlike termites, ants are not famed for being particularly skillful builders, and even the most imposing structures of our indigenous species are dismissed as 'ant hills'.

In fact, however, the ant hills of our indigenous ants have a plethora of internal subdivisions and conceal their inhabitants in an ordered structure of chambers and passageways. They represent intelligent and technically accurate constructions which, in their position and structure, fulfil the various temperature and moisture conditions necessary for the development of the brood. A number of species can also, to a certain extent, directly influence the temperature and moisture levels inside the nest. This occurs in the case of the wood ants by workers warming themselves in the sun on top of the ant hill in the springtime ('sunning period'), and then giving off the stored heat like living ovens inside the nest. Some species regulate the moisture level of the nest by bringing in water, such as, for example, *Monomorium pharaonis* (Sudd, 1960, 1962), which is able by this means to live in nests of 58-66% rel. atmospheric moisture (Peacock *et al.,* 1955).

12.1 Ants without nests

Army and driver ants form the main group of ants which lack permanent nests, and these ants are grouped together in the Doryline subfamily. Their life cycle comprises two distinct phases, the nomadic and the stationary phases. During the nomadic phase, the colonies change part of their camping place on a daily basis. They travel further during the day when they also bring in large amounts of prey, the greater proportion of which they feed to their brood in the old camp. As darkness begins to fall, a new camp is frequently made; the old bivouac is dissolved and the workers which had remained behind move with queen and brood into the new domicile. This marching activity during the nomadic phase, which guarantees an adequate supply of food for the numerous offspring, is very probably maintained by means of the secretion from the labial glands of the larvae (Wang and Happ, 1974). This secretion is no longer produced when the larvae pupate, and simultaneously the food requirements of the colony are greatly reduced. The marching activity of the army ants ceases when the larvae pupate and the stationary phase begins, during which time the bivouacs are maintained for longer periods. The food brought into the colony in the stationary phase is mainly for the benefit of the queen. Her rear segment swells into a shapeless mass, becoming to some extent physogastric, a characteristic unique to ants, and the queen soon begins to lay eggs. The young larvae hatch from the eggs at about the same time that young workers emerge from the larvae of the previous generation. With the hatching of the young workers, which is stimulated by a pheromone from the freshly hatched animals, the raiding activity begins anew (Topoff, 1972), thus marking the start of a new migratory phase. This is once again maintained by the secretion from the labial glands of the newly emerged larvae and the cycle begins again (Fig. 12.1). The characteristic features of this cycle were discovered by Schneirla, who as a psychologist, investigated the behaviour of South American army ants both in the laboratory and under natural conditions, from 1932 until his death in 1968.

Before we look more closely at the structure of the army ant bivouacs, we must first distinguish between two different streams of development within the Dorylines, between those adapted to life on the surface, the epigeic species, and the hypogeic species. Those belonging to the epigeic species are in particular species of the *Eciton, Aenictus* and – less obviously – the *Neivamyrmex* genera. The *Dorylus, Anomma, Labidus* and *Nomamyrmex* species are hypogeic. *Eciton, Neivamyrmex, Labidus* and *Nomamyrmex* occur only in southern and central America; the remaining genera are representatives of the Old World, living therefore in Africa, south east Asia and Australia. We shall deal next with the conditions prevailing among the epigeic species, that is as they are most typically expressed among *Eciton* and *Aenictus.*

A bivouac is not a nest in the usual sense of the term, built by the

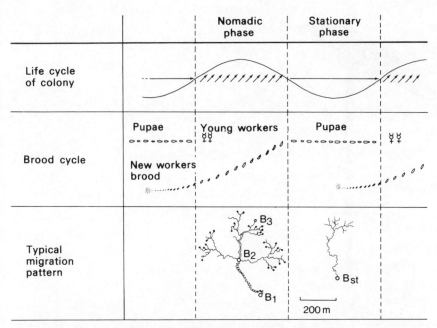

Fig. 12.1   Various phases in the behaviour of the army ant *Eciton hamatum*.
B. bivouac places of the nomadic (B_{1-3}) and the stationary phase (B_{st}), arrows:
extended movement during the day and migration during the night (after
Schneirla, 1971).

ants in the ground or another substratum; rather it is formed with living
nest mates which join together in a suitable place. This joining together
of the army ants is a reaction to stimuli coming from prey and occurs
when daylight fades and the temperature drops. The army slows down
as stimuli from their own colony, such as trail substance, colony scent
and tactile stimuli from colony members predominate. The ants 'lump'
together, taking hold of each other by the legs which can clasp
effectively due to the size and shape of the claws. From one such clump
of ants, a hanging chain of ants usually emerges (Fig. 12.1), soon
becoming denser, binding more and more members of the colony. The
bivouac slowly forms; other bivouacs are started simultaneously in a
number of places, until they eventually become one.

   A completed bivouac is a cluster of living ants, enclosing the colony
within itself. It is orderly to a high degree. The older and more robust
ants are gathered on the periphery, the younger ones inside. The ants
surround the queen and the entire brood, which are able to live inside

Fig. 12.2   Chain formation of a laboratory colony of the army ant *Eciton hamatum*. In natural conditions this leads to the formation of a bivouac (after Schneirla, 1971, adapted).

owing to openings and passageways in places where the texture of living ants is separated. The busy comings and goings of ants laden with brood or prey usually prevail along these passages. At the beginning of the nomadic phase, the inside of the bivouac is also strictly divided into sections. The queen, brood and prey are all separated into different chambers, connected by passages. Towards the end of the nomadic phase, the bivouacs become more voluminous and less orderly and compact.

In this fascinating work of cooperation, a population of several million animals fulfils four objectives in the bivouac. They serve as the basis and centre for colony operations; as shelter and protection; as incubator for the brood; and as a reservoir for the population. These four functions are valid mainly for the bivouacs of the epigeic species, and in particular for *Eciton* (Schneirla, 1971).

The bivouacs serve as the centre base for the operations of the colony mainly during the day, when the workers swarm out of the colony in different directions to catch prey. As darkness begins to fall, a new bivouac is set up in another place and the removal of the queen and brood organized.

During the night and the following day the bivouacs serve the colony, or a part of it, as shelter. For those sheltered in the bivouac, and for those forming it, the important question is that of maintaining the temperature and moisture of the bivouac at a tolerable level. This is accomplished usually by the tendency for the bivouacs to be set up in moist, shady places. If it should turn out on the following day, however, that the bivouac is in the sun, then the ants may evade it. The animals on which the sun is shining directly become uncomfortable, break their chain and rejoin the bivouac in a shadier spot. If by such readjustments the animals are still unable to avoid the sun, the bivouac may dissolve completely and reform in another place.

Smaller climatic swings inside the bivouac are regulated in a different way. If, for example, the external temperature drops to such an extent that the temperature inside the bivouac threatens to fall below a critical level, then the entrances to the bivouac are closed and the outside walls thickened by workers from the inside, so that more heat is kept inside the bivouac. With a rising outside temperature on the other hand, workers disengage themselves from the outside walls, thus rendering them more porous, so that the circulation of air inside the colony is increased. This has the effect of causing more moisture inside the bivouac to evaporate and thus reduce the temperature in relation to the outside. By these means, the temperature achieved and which prevails inside the bivouac varies less than that in the surrounding environment. In the tropical forests in which *Eciton* lives, for example, the temperatures in the area surrounding the bivouac vary daily from about 22°C to 30°C, whilst the temperature inside the bivouac has a maximum variation of only 3°C.

The bivouacs of epigeic army ants can be made in the most widely differing places and assume a variety of forms (Fig. 12.3). For example *Eciton burchelli* mostly bivouacs in the form of clusters in tree branches up to 30 m from the ground. *Eciton hamatum,* on the other hand, does not bivouac more than 1 m above the ground. The form of this bivouac frequently resembles a curtain falling diagonally along the root of a tree,

or between a branch and the ground. Other *Eciton* species, such as
Eciton mexicanus, Eciton vagans and *Eciton dulcis,* seek darker and

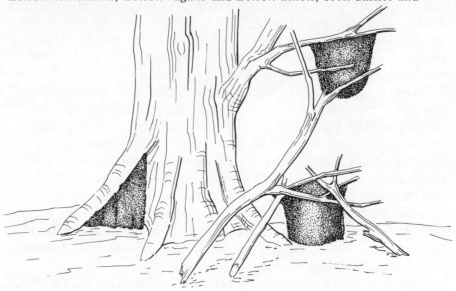

Fig. 12.3   Various bivouac forms of the *Eciton* genus of army ants (after Sudd,
1976b — from Schneirla, Brown and Brown, 1954, adapted).

damper spots for their bivouacs (Schneirla, 1947; Borgmeier, 1955;
Rettenmeyer, 1963). The bivouacs of the stationary phase are placed in
more protected, and usually underground spots — with the exception of
Eciton burchelli and *Eciton hamatum.* Among the *Aenictus* genus, the
spectrum reaches from obviously above-ground species, such as *Aenictus
laeviceps* and *Aenictus gracilis,* to species which live more or less
underground, such as *Aenictus aratus,* which also place their bivouacs in
the nomadic phase under stones or in holes in the ground (Schneirla and
Reyes, 1966).

The expressly hypogeic species differ in a number of respects from
their epigeic relatives. Corresponding to their underground lifestyle, the
eyes of epigeic ants are less effective than those of the species living
above ground, and may have no function at all. The bivouacs are all
placed underground at depths of up to 4 m, sometimes also in holes in
the ground which they have dug out themselves, such as among
Anomma (Raignier and von Boven, 1955). With their mostly
underground lifestyle, the hypogeic species are not able — but in any

case do not need — to actively regulate the climatic conditions inside their bivouacs.

12.2 Ground nests

Ground nests are found quite frequently among ants, especially in the Ponerine subfamily, and count as an original form of habitation, from the beginning of their evolution. This is supported by the observation that the more primitive Ponerine genera, such as *Ectatomma, Paraponera, Harpegnathus, Dinoponera, Streblognathus, Megaponera* and *Odontoponera* and also the majority of Myrmicines, nest in the ground. Additional evidence to support this is the fact that only those ants belonging to the Ponerine complex are able to stridulate, and thus have at their disposal a communication system which we know, at least among the leaf-cutting species, serves as an underground alarm system (Markl, 1973a). Whether or not all ants stem from the ground dwellers is uncertain. According to one theory of Malyshev (1966), ants originated not from ground, but from tree-dwellers (*see below*).

The ground nests of ants are not at all homogeneous, displaying quite substantial differences between only slightly structured nests, ranging to the complicated structures of European red ants or the South American leaf-cutter ants. Among some species, the nests are placed in the ground, and then transplanted into rotten wood, as may occasionally be observed with our *Lasius niger* species, amongst which, however, the reverse may also occur as it does with carpenter ants, where the nests are begun in the wood and the constructions completed in the ground (*see*, for example Eidmann, 1928b).

Ground nests of very simple structure may be found, for example, among the Australian species *Myrmecia dispar* (Gray, 1971a). In a young colony, the nest consists of a narrow passage, leading into the ground almost vertically and widening out into a brood chamber at its base. Close to the ground surface, a side passage leads off which in fact leads nowhere and is filled with the waste from the colony (Fig. 12.4a). As the young colony grows in size, the nest gains in depth; the main passageway, which reached originally 15 cm into the ground, later reaches about 80 cm. In addition, further side passages are dug out which can lead in all directions. In this way, nests as depicted in Fig. 12.4 eventually emerge, but which in their essential elements continue to resemble the founding nests. In both cases, the brood and the queen are to be found at the deepest point in the nest and thus also the safest

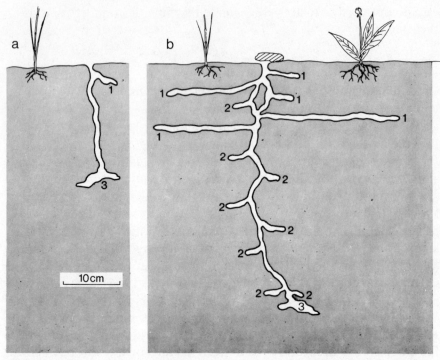

Fig. 12.4   Nests of *Myrmecia dispar;* (a) founding nest; (b) nest of an established colony; 1. waste chambers; 2. nest chambers; 3. main brood chamber (after Gray, 1971, adapted).

and most moist place, which is particularly important in desert and steppe areas. Only in times of great reproduction in the colony are the higher chambers used for keeping the brood. The waste from the colony, formed from pupal cocoons, dead nest mates and the inedible remains of prey, are collected in the least safe and at the same time least habitable side passages, 3-20 cm under the surface. The entrances to the nests are usually concealed under twigs, leaves or clumps of grass; because of this, and because of the underground storage of nest wastage, the nests are difficult to see from the outside.

Similar nests, with passageways placed vertically into the ground, are constructed by ants from more highly evolved subfamilies, such as *Formica fusca* found in central Europe, *Prenolepis imparis* indigenous to the Mediterranean area, the North American species *Pogonomyrmex badius* and various corn ants, such as *Messor* species and also *Veromessor pergandei*. The nests of *Formica fusca* consist of numerous vertical passages, leading down into the ground up to 70 cm; small brood cells lead off from these passageways (Brian and Downing, 1958). Substantially deeper than the *Formica fusca* nests are those of *Pogonomyrmex badius,* which may be over 2 m deep (Wray, 1938). The nests of *Veromessor pergandei* attain depths of 3 m and those of *Pogonomyrmex barbatus* even depths of 5m (Whitford *et al.,* 1976).

In contrast to these nests, which mainly lead downwards into the ground, are the nests of *Lasius alienus* and *Tetramorium caespitum,* which are more horizontally oriented. Neither of these European species requires a substantial nest depth in order to achieve the necessary moisture level. Dietary habits determine the structure of these nests. *Lasius alienus* and *Tetramorium caespitum* live to a large extent on the excreta of root lice. The nest passages of these species accordingly lead more in a horizontal direction, where root aphids are to be found. Other species, such as, for example, *Crematogaster auberti* (Fig. 12.5), follow the tree roots with their passages, thus reaching the root lice which are essential for their diet (Soulié, 1961).

Obviously in the construction of the ground nests, earth has to be brought out of the inside of the nest and off-loaded. Among *Myrmecia dispar,* the substratum brought out is distributed at some distance from the entrance to the nest, in such a manner that no hill is formed and the entrance to the nest is not noticeable (Gray, 1974). Other species lay the superfluous earth at the entrance to the nest, thus forming a crater, as has been described for the nests of *Cataglyphis bicolor* in North Africa and the Near East (Wehner and Lutz, 1969; Wehner, 1970), *Cataglyphis cursor* (Parashivescu, 1967), *Camponotus compressus* and *Messor arenarius* (Wehner, 1970; Delye, 1971) (Fig. 12.6). Crater nests like these are particularly known in desert and steppe regions. Again, other species make use of the refuse from their nests for the construction of a hill above the entrance, this being fortified by other material from the vicinity of the nest, and containing passages and chambers within it. *Cataglyphis bicolor* belongs to those species which build either crater nests or hill nests according to the situation of their nest. The crater nests of this species are thus found very widely, whilst the hill nests have only been found in Afghanistan, and there in fact in areas with coarse-grained soil conditions and marked drops in temperature from day to night (Schneider, 1971).

Among *Messor ebenius,* a harvester ant of the Sahel area, Tohme (1972) found nests dug deep into the ground, consisting of two parts, one lying on top of the other. Both parts contained passages and

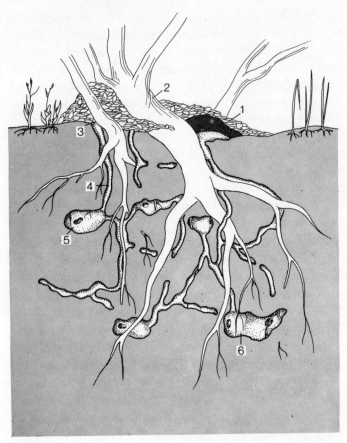

Fig. 12.5 Nest of *Crematogaster auberti*. 1. Nest refuse and main entrance to the nest; 2. tree, *Passerina hirsuta;* 3. secondary entrance to the nest; 4. nest passage; 5. brood chamber; 6. nest chamber in which the queen is kept (after Soulié, 1971, adapted).

chambers, but they served different purposes: the upper section of the nest, which dried out completely in the summer, served as a corn store, whilst the deeper section reached into the moister areas of the ground and was used as accommodation for the colony.

Given the various different kinds of earth nests which are built by ants, it is to be expected that the ants will correspondingly orient themselves in different ways whilst digging the passages and chambers. Sudd investigated the orientation of individual species during digging, in the laboratory (1967a, 1970a,b) and determined that there were in fact corresponding differences. Passages leading in a perpendicular direction downwards were bored by *Formica lemani* in a straight line, but not by

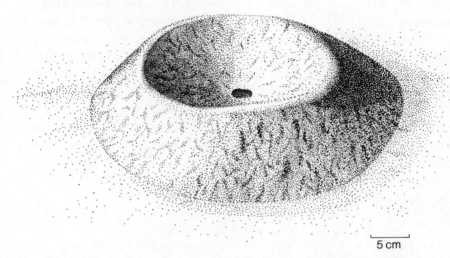

Fig. 12.6   Crater nest of *Messor arenarius* in Tunisia (after Wehner, 1970).

Lasius niger workers, which dug the tunnel in a more horizontal
direction as they continued. The passages of *Lasius niger* workers
broadened increasingly and forked off in many places. The passages of
scaled ants, such as, for example, *Myrmica scabrinodis,* showed the same
characteristics as those of *Lasius niger,* but much less markedly
expressed. The passages dug by *Formica lemani* and *Formica fusca,* on
the other hand, proceeded into the ground completely vertically, in a
straight line and completely without any branchings off.

   Many ants in temperate latitudes build ground nests with hills. The
advantage of nest hills is that energy from the sun can be used more
effectively for warming the nest than nests without hills. According to
research conducted by Steiner (1929), in the morning and evening
hours a nest hill receives three times as many rays from the sun as a flat
piece of ground with the same diameter. The temperature underneath
the top of the nest hill in the case of the nests of *Lasius flavus, Lasius
niger* and *Formica fusca* is 3-7°C above the temperature of the
surrounding ground in the daytime (Cloudsley-Thompson and Sankey,
1958; Steiner, 1929). Steiner reached similar conclusions (1924) in his
investigation of wood ant nests. During the brood period, he found an
average dome temperature at a depth of 30 cm of close to 26°C, which
was 10°C above the average ground temperature of about 16°C. In the
case of *Formica pratensis,* Galle (1973) found in the time between 29
May to 1 June an average temperature of approaching 30°C in the nest

hill, whilst the ground temperature was slightly above 16°C and the temperature of the air at ground level was about 19°C. This temperature range of 23-29°C exactly corresponds to the optimal temperature for ants, as was shown, for example, by the investigations of Herter (1923). Inside the nest the same temperature does not, however, prevail in all places, and even taking atmospheric moisture into account, there is a tendency for the temperature to go counter to the optimum. These varying climatic conditions inside the nest are nevertheless favourable to the inhabitants; they provide the ants with the possibility of selecting what for them is the optimal from out of the spectrum of various conditions available. For the brood, such differing conditions prove to be a necessity, since the pupae are particularly in need of warmth, but are relatively resistant to dryness, whilst eggs and young larvae require less warmth, but more moisture (Janet, 1904).

From another point of view, a hill is more inclined to be carried off by the wind and correspondingly cools off more quickly without sunlight, than does flat ground (Steiner, 1929). Among *Lasius niger* and *Tetramorium caespitum,* it has thus also been observed that the ants only live in their nest hill in the summer, withdrawing deeper into the ground in winter whilst the wind and weather destroy the hill, which is reconstructed the following spring (Donisthorp, 1927).

In tropical regions, ant hills are obviously not necessary in order to make use of sunlight and for warming the nest. This does not result simply from the fact that species indigenous to these areas manage without ant hills. Temperature measurements taken by Levieux (1972) in the ground nests of *Camponotus acvapimensis* on the Ivory Coast show that temperatures here are relatively high and variations minimal. In one nest situated in a savannah region, the temperature at a depth of 30 cm averaged 26°C, involving a daily, and even a yearly variation of only about 2°C. The relative moisture level here was 100% throughout the entire year.

The most imposing ant hills are built by our wood ants, these having been reported to be over 1 m in height and 9 m in circumference (Stammer, 1938); Wasmann (1934) even reported one ant hill as being 17 m in circumference. The chambers inhabited by the ants are situated not only in the visible hill itself, but may in fact reach as far below ground level as the hill towers above it. Such constructions are begun either in the ground itself, where a number of hollows are made first, or alternatively in a tree stump, in the rotting wood of which the first few nest chambers are made. In the construction of the dome of the nest, the bits of plants which have been procured for the purpose are not just piled on top of one another indiscriminately, but placed systematically. The coarser material is arranged on the inside of the ant hill, the finer material being placed around it to form the 'nest mantle'. The dome owes its stability to this coarser nest material, whilst the fine building material at the periphery of the nest is so densely packed that it effectively protects against moisture and loss of heat (Gösswald, 1951; Kloft, 1959b). The entrances left in the nest mantle are closed during the night and opened up again during the day. The 'doors' are kept closed in cold weather and are correspondingly opened wider in hot conditions.

The shape of the ant hill may vary between the extremes of a completely flat nest and a steep dome-shaped nest, according to what is available in the area and the exigencies of heat, and probably also moisture, in the local environment. There are besides this scale of extremes also unusual constructions, such as nests with two peaks, nose-like protrusions, or a series of smaller hills before the nest proper. In sandy soils, one occasionally finds nests where the hill itself is below the ground surface. In purely deciduous forests nests are not usually particularly large, although the colonies may be extraordinarily large (Lange, 1959). The shape of the ant hill, in particular the steepness of the dome, may be substantially affected by external factors, as emerges from the experiments conducted by Lange (1959). In the laboratory, Lange warmed wood ants by means of infrared lamps, or cooled them with cooling troughs using water. This revealed that the ant hills became flatter with increasing external temperature and steeper with decreasing external temperature. Moisture also has an influence on the shape of the ant hill as shown by the experiments of Zahn (1958), who was able to demonstrate that the building activity of the ants increased after the nest had been dampened. The greater the level of moisture in the nest, the steeper were the nest embankments.

These experiments make it abundantly clear that the ant hills of the wood ants are not inflexible structures, but may be completely rebuilt and altered. This is demonstrated particularly forcefully by Kloft's experiments (1959b), where the dome of a *Formica polyctena* nest was sprayed with a blue dye, which thoroughly permeated the nest material without sticking it together. After a number of days the ant hill still had its former appearance, but the coloured portion lay 8-10 cm below the surface. The nest was then sprayed with yellow dye, and since this also became buried below the surface of the hill, it was then sprayed with red dye. Four weeks later, the colours reappeared on the surface in their original order. This experiment shows that the ants are continually bringing nest material out to the surface of the nest from inside, whilst the surface layer correspondingly becomes buried inside the structure, until it too is brought out to the surface once again. In this way, the material from the moist inside of the nest is dried on the surface, and mould formation is avoided. One can see how important this process is very clearly from abandoned nests, which are very quickly penetrated by mould.

Lasius niger, Lasius alienus and the yellow field ant *Lasius flavus* erect far smaller constructions. They do not collect any plant material for the nest, constructing their hills only out of earth brought up from the deeper areas of the nest and gathered from around the nest site. The inside of the nest contains chambers and passages in numerous levels, one on top of the other. Each level is about 1 cm high and the walls between levels are often only 1 mm thick. Such an ant hill built from porous earth is not very stable and may frequently be destroyed by wind and rain. These hills, however, would be far more unstable if they were not — as is usually the case — built around plants, which act as columns around which construction can proceed (Fig. 12.7). Far more stable than these ant hills are those of the yellow field ants, which are substantially more densely packed and have plants growing in the structure. These plants take root in the ant hill and give it quite considerable resilience.

Ants which nest under stones enjoy advantages similar to those resulting from the construction of ant hills. Various *Lasius, Myrmica* and *Leptothorax* species belong to those ants found under stones, in addition to which are *Tetramorium caespitum* and various other species. As with ant hills, the temperature of a nest under stones when the sun is shining is often considerably higher than that of the surrounding ground, as is shown in the measurements taken by Steiner (1929). Here, however, the thickness and size of the stones play a considerable part.

Three main factors are important in order for the underside of the stone to become warmer: a greater absorption of heat by the stone due to its projection above the surrounding ground surface, a greater heat conductability of the stone in relation to the ground and finally a higher heat capacity of the stone in relation to the ground, which is chiefly dependent on moisture. A slower heat loss and thus also a slower heat storage by the stone in relation to the ground may not however, be observed and also goes against its relatively good heat conductability.

Among ant species which construct dome nests as well as nesting under stones, it is striking that in higher mountain sites they nest almost exclusively under stones. This is very probably connected with the fact that above a certain height the level of moisture of the ground is higher in mountainous areas due to rainfall and mist, and the stones therefore lose less heat because of their faster drying out. In addition, in these high areas frosty nights occur even in summer. In this case a dome nest loses a substantial amount of heat whilst thawing out, whereas the temperature of a stone increases immediately at the onset of sunlight.

Fig. 12.7   Ant hill of *Lasius niger* (after von Frisch, 1974, adapted).

Finally, stronger and more frequent winds occur in mountainous regions, and these would reduce the heat from a dome nest more than that of a nest under a stone (Steiner, 1929).

The largest ground nests built by ants are those constructed by
leaf-cutter ants in South America. Some of their constructions are over
4 m deep and can penetrate as much as 8 m of ground with passages and
numerous chambers. These nests were investigated mainly by Eidmann
(1932) and Jacoby (1953, 1955), who poured cement into the nest
openings and, after detaching the cement and removing the earth,
obtained a passage system of the nest. This passage required about
1.5 m³ of cement.

The substantial dimensions of these nests are not only necessitated
by the large number of their inhabitants, but also by their particular
feeding habits. Leaf-cutter ants feed exclusively on fungi, which they
cultivate in their enormous nests on chewed up vegetation. Not only do
these fungal gardens require space, but they also necessitate certain
conditions of culture, which again determine the architecture of the
nests.

The structural plan of an *Atta sexdens rubropilosa* nest is presented
in Fig. 12.8. The entrances to the nest lead into passages which proceed

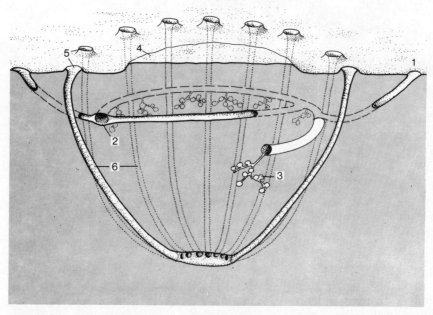

Fig. 12.8   Nest of the leaf-cutter ant *Atta sexdens*. 1. Nest entrance; 2. cyclical
channel; 3. brood chambers and fungus garden; 4. nest refuse; 5. crater; 6. boundary
channel (after Jacoby, 1953, 1955 and Sudd, 1967b, adapted).

quite evenly, hitting a cyclical channel at a depth of 60-80 cm, the channel lying horizontally and having a diameter of exactly 1 m. This cyclical channel serves amongst other things for the traffic of in-coming and out-going ants and to maintain the easy transportation of leaf material. The passages to individual fungus chambers lead off from the cyclical channel; these chambers show a diameter of 24-30 cm, and are so densely packed together that the walls dividing them are only paper thin.

Given such dense packing together of the chambers, the fungi in the ground nests are only able to thrive when the ants ensure that the heat produced by the fungi and the carbon dioxide which they release are removed, at the same time providing sufficient oxygen. This occurs by means of a second system of passages connected with craters, which form a circle around the central point of the nest. So-called boundary channels lead from these craters at an agle of about $50°$ into the ground, re-joining at the deepest point of the nest. Oxygen measurements showed that these boundary channels in fact ensure that an exchange of gases takes place. Here the diagonal path of the passages plays an important part. The fresh cool air proceeds along the length of the boundary channels, pushes its way up into the passages and chambers, thus displacing the used warm air. This exchange of used air for fresh cool air takes place as a rule during the night.

12.3 Wood and 'carton' nests

According to a hypothesis of Malyshev (1966), the colonies of the oldest ants and their ancestors were constructed in decaying wood, where the queen and her first young fed themselves on the fungi growing there. With the growing colony, space as well as food became scarce and the young workers expanded the nest by building passages and at the same time finding fresh food sources. Soon the diet would consist not only of fungi, but also of insect larvae living in the wood. Larger pieces of prey would either have to be transported to the larvae in little pieces, or else the larvae transported to the prey. The observation that today still, among primitive ant species, e.g. Amblyopone, the larvae are placed on to the prey is regarded as an indication of the accuracy of this hypothesis. If, however, one compares this hypothesis with the nesting habits of the ants living today and with the structure of the stridulation organ (*see above*), then it seems unlikely that this hypothesis holds for all ants in general.

Belonging to the wood-inhabiting ant species are, first, the small *Leptothorax* species, which gnaw out the dead branches of trees for their poorly populated colonies until they have sufficient space. Other *Leptothorax* species prefer the trunks of trees, frequently living under the bark and feeding on beetles in particular. They may also be found in tree stumps, however, and also in small dead branches strewn on the ground in woods.

One group of ant species lives in trees without nesting in the wood. They settle in the protruberances of deciduous trees, obtaining ready-made habitation from the tree, the production with which, however, they are not involved. Torossian, who investigated the fauna in gallnuts, found several ant species, mainly *Leptothorax rabaudi*, *Leptothorax nylanderi*, *Dolichoderus quadripunctatus* and *Colobopsis trucatus,* and amongst them also species which are also able to nest in other places (Torossian, 1971a,b, 1972).

The species named so far settle only in the peripheral areas of trees, if not in broken-off branches or under stones. This is not the case among *Lasius* species, such as *Lasius brunneus*, whose very populous colonies may also inhabit, apart from many other places, the insides of trees (Buschinger and Kloft, 1969), nor is it at all the case among carpenter ants. Carpenter ants are the largest of the central European ants, being chiefly represented in Germany by the species *Camponotus herculeanus* and *Camponotus ligniperda. Camponotus herculeanus* nests are mostly to be found in living trees, although occasionally in dead wood. *Camponotus ligniperda* inhabits the same nest areas as *Camponotus herculeanus,* but frequently makes do — like *Camponotus vagus,* which is prevalent in warmer regions (Benois, 1972b) — with dead wood. The nests of the carpenter ants follow the year-rings in the wood with their sites. These year-rings consist of a light-coloured, relatively soft part, which emerges during the rapid growth of spring, and a harder part, the somewhat darker summer wood. The ants prefer to gnaw out the softer portions, which is easier for them and in addition secures for them the advantage of more stable nests. In this manner, apparently healthy-looking trees are gnawed out to a height of some metres and inhabited by the ants.

The nest area size of individual *Camponotus herculeanus* nests was ascertained by Kloft *et al.* (1965) by means of tracer experiments. They fed the ants with radioactive iodine in hydromel and were able — after the food had been distributed throughout the colony — to determine the boundaries of the nest area by means of the radioactivity. They found that the nests of individual colonies could include up to 12 trees and extend over a surface area of 130 m². The individual nesting trees are connected with each other by passages leading along the root systems of the trees. Only small intermediary stretches lead through the surrounding earth. Within a circumference of 2 m around the base of the trunk, the main roots were particularly densely occupied by ants.

The shiny black wood ant *Lasius fuliginosus* prevalent in central Europe is also counted among the tree-inhabiting species. In contrast with the carpenter ants, however, it does not create its living space by gnawing out the wood, but by building its nests into already existing hollows. Its nests are found mostly at the base of living deciduous trees, the heartwood of which has been destroyed by fungal attack; more rarely, they may be found in old tree stumps or other places. The ants build their nests in these hollows, the nests consisting of irregularly ordered chambers made from a dark substance similar to carton. The ants usually only remain in these chambers during the summer; the majority of them spend the winter in a deep-laid ground nest (Eidmann, 1943), whilst a smaller proportion of the colony may remain behind in the summer nest (Maschwitz and Hölldobler, 1970).

The 'carton' material from which the wood ants build their nest chambers, consists mainly of fine sawdust, held together with a cementing material. By way of cementing substance, the ants do not use, as was previously supposed, their mandible gland secretion, but sugar solution, with which they cultivate a special fungus. In central Europe this occurs only in the building material of *Lasius fuliginosus* (*Cladotrichum myrmecophilum,* according to Lagerheim, 1900). The *Lasius fuliginosus* nests owe their stability to this fungus, which is evidently not eaten by the ants (Maschwitz and Hölldobler, 1970). Similar 'carton' nests are also found amongst other Lasius species, such as *Lasius emerginatus* (Wasmann, 1913), *Lasius rabaudi*, whose carton nests are covered with attractive, velvety fungal myceli (Kutter, 1969), and among *Lasius umbratus* (Adlerz and Bönner, after Stitz, 1939).

The *Crematogaster* genus includes a variety of species, distributed across a wide area from the Mediterranean to the tropics. They construct nesting places not only in the ground as does, for example, *Crematogaster auberti,* but also in wood (e.g. *Crematogaster scutellaris*), the nests being penetrated by a large number of passages. They are also able to subdivide tree hollows with carton nests (Soulié, 1961; Casewitz-Weulersse, 1972). A number of tropical species place their nests outside trees altogether, constructing nests made entirely of carton material, which may occasionally hang as much as 1 m from a tree, as has been reported for *Crematogaster depressa* in Africa (Strickland, 1951), as well as for *Crematogaster skouensis* in south east Asia (Soulié, 1961). Carton nests of this kind hanging from trees are also found among other genera, such as *Nematocrema,* which is found in Africa (Delarge-Darchen, 1972b; Fig. 12.9). It is still not known of what materials these carton nests consist. Soulié in fact assumes that

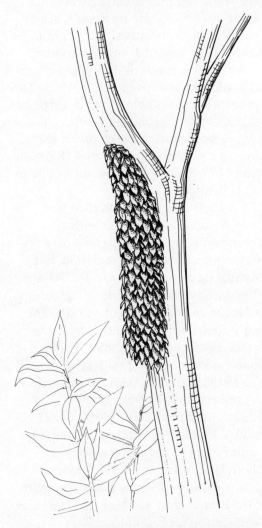

Fig. 12.9   'Carton' nest of *Nematocrema* (after Delage-Darchen, adapted).

they are produced from wood material and mandible gland secretion, but this hypothesis remains uncertain, as no specific research on the subject exists.

12.4 Nests in ant plants

Ant plants are inhabited by ants particularly frequently, since they offer especially suitable nesting places with their ready-made hollows. Among those ants which obtain a place to nest in these plants, there is a large number of species which are not in fact dependent on these ant plants, or myrmecophytes, such as *Crematogaster, Tetraponera, Monomorium* and others. They nest wherever suitable hollows are available and thus also claim those provided by the ant plants. Other species, on the other hand, especially the Pseudomyrmecine subfamily, but also the Dolichoderines (particularly *Azteca*), the Myrmicines and the Formicines, only nest in specific plants and are necessarily dependent on their host plants. They are thus described as obligatory plant-inhabiting species.

But what is the situation with the plants themselves? Are they dependent on colonization by ants, in the same way that certain ant species in their turn need the plants? Or do the plants in some way gain an advantage through being colonized by ants, whereby the formation of the hollows inhabited by the ants (myrmecodomati) has been selectively encouraged in the evolution of the plants? It is this question – whether or not a coevolution has taken place between plants and ants – which has played a part in defining the present manifestations of plants and ants.

Around the turn of the century, this hypothesis was the stimulus for a large number of works about myrmecophytes (*see*, for example, Warburg, 1892). Since, however, no overwhelming indications of a coevolution between specific ant species and their nesting plants could be found, and since instead of this only advantages for the ants could be discovered, there being no real gain for the plants offering them shelter, interest in the ant plants and their relationship to the ants waned. Only later, mainly due to the work of Janzen (1966, 1967a,b, 1969, 1972, 1973, 1975) did this interest revive and increasingly reveal a mutual advantage in the relationship for plants and ants.

For the majority of myrmecophytes known so far, however, the advantages to be gained through colonization by ants have still not been clearly proven. This is the case, for example, for numerous leaf domati

known in tropical South America (Bequaert, 1922; Wheeler and
Bequaert, 1929; Vogel[1]).

Among the *Tococa* genus, which belongs to the Melastomataceae, for
example, pocket-like structures of differing forms[1] emerge from the
leaves, the structures also being in pairs, each containing an opening on
the underside of the leaf (Fig. 12.10). The pockets are situated usually

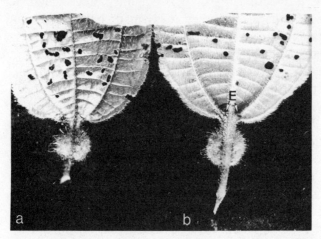

Fig. 12.10   Tococa guayanensis with Myrmecodomati. (a) upperside of leaf;
(b) underside of leaf with entrance openings (E) into the domati (Photo —
St Vogel).

at the base of the leaves and are definitely not galls, since they are
formed by the plants themselves without any action on the part of
gall-flies or ants. The shape and situation of the pockets are
characteristic for different *Tococa* species. The inside of the pockets is
hollow, and partially subdivided into a number of chambers[1] .

These pocket-like formations of the *Tococa* genus are frequently
inhabited by ants, which in certain cases specialize in specific *Tococa*
species. The various chambers inside some of these Myrmecodomati
are filled with the larvae. In some species, the individual chambers are
supplied with separate exit passages, behind which sentries frequently
stand[1] . Among the Myrmecodomati there are also some which are
equipped with a large number of little hooks inside: these are used by
the ants, amongst other things, for hanging up their larvae (Fig. 12.11c).
Of the pair of domati, the ants usually inhabit only one. The other
often serves as a 'louse stall'[1] , in which Pseodococcides (Coccidae

1. From a lecture given by Professor St Vogel in Kiel in February, 1975.

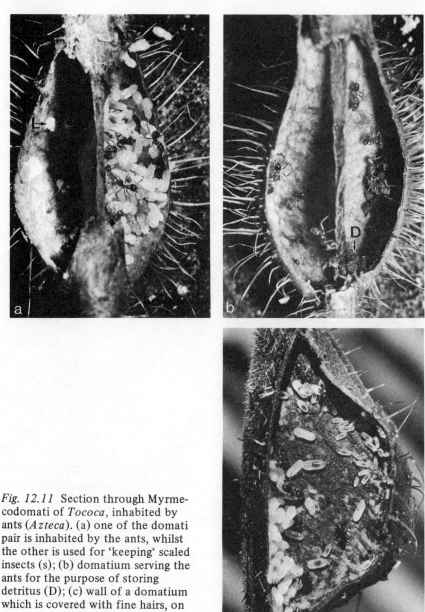

Fig. 12.11 Section through Myrmecodomati of *Tococa*, inhabited by ants (*Azteca*). (a) one of the domati pair is inhabited by the ants, whilst the other is used for 'keeping' scaled insects (s); (b) domatium serving the ants for the purpose of storing detritus (D); (c) wall of a domatium which is covered with fine hairs, on which the ants hang the larvae (Photo – St Vogel).

family) tend to be kept, the sweet excretions of which the ants live on
(Fig. 12.11a). Occasionally insects are also killed, thus not only
obtaining necessary protein food, but also ensuring that the insects do
not get out of hand. In addition to domati which are used as
accommodation for plant lice, others are also known to be used by the
ants as 'latrines' and refuse chambers (Fig. 12.11b).

Apart from the laminal domati of the Melastomataceae (e.g. *Tococa*),
there are also ligular domati (among Euphorbiaceae), stipular domati
(among certain acacia species of Mexico), rhachis domati (among
Leguminosae) and, finally, shoot and root domati. Plant families from
which Myrmecodomati are formed are to be found, in addition to
Melastomataceae, Euphorbiaceae and Leguminosae, among Rubiaceae,
Borraginaceae, Moraceae and Sterculiaceae.

The advantages for the plants being colonized by ants probably lie in
the fact that they obtain food from the detritus of the ants. This is
supported by the researches of Benzing (1970) with the Myrmecophytes
Tillandsia butzii and *Tillandsia caput-medusae,* which belong to the
Bromeliaceae. It is possible that apart from the advantage of food the
ants also keep harmful fungi away from the plants, thus serving the
hygiene of the host plants.

On the Malaysian archipelago, plants of the *Myrmecodia* and
Hydenophytum genera live as epiphytes on other plants. They produce
nodular structures containing a spongy tissue rich in sap, as well as
numerous irregular hollows. These hollows are frequently inhabited by
ants of the *Iridomyrmex* genus, which use some of the chambers as
living space and others as latrines, thus possibly fertilizing their host
plants.

In the case of ant gardens, the advantages for the plants and the ants
are easier to see, although no myrmecodomati are constructed from
these plants. For example, without ants *Codonanthe uleana* can only
flourish in the soil and not on other plants. This plant, however, grows
anchoring roots, with which it attaches itself to the roots of plants.
Various ant species, particularly species of the *Azteca* and *Camponotus*,
are then able to add to these anchoring roots by bringing earth and
placing it among the roots. The plants take root in this earth, thus
holding it together. In this way, structures of substantial size may
eventually emerge from this earth in which plants and fungi have
grown; these serve the ants as nests, and enable plants such as
Codonanthe to live epiphytically on the stems of other plants
(Fig. 12.12; Ule, 1905, 1906). Other plants which, together with ants,

form 'ant gardens' are those belonging to the *Hoya, Dischidia* and *Aeschynanthus* genera, which were investigated by Leeuwen (1929) in Java.

Fig. 12.12 Ant garden. The plant *Codonanthe* fastens itself to the trunk of a tree with its anchoring roots, and roots itself in the earth placed there by ants (*Azteca* species). The completed ant garden has the advantage for the *Codonanthe* of enabling it to live epiphytically and for the ants (*Azteca*) of providing a stable nest by means of the *Codonanthe* roots (Photo — St Vogel).

One of the best researched examples of ant plants is *Cecropia adenopus,* which grows in the primeval South American forest. Their slender trunks, which carry three-cornered leaf scars, grow to heights of 12-15 m, forming quite small crowns with branches arranged in the form of candelabra. Large serrated leaves are found on the ends of the branches, the leaves having thorny petioles. The insides of the stems and twigs are hollow, and occasionally have partitions formed at the nodes, thus dividing the hollows of the stems and branches into individual chambers (Fig. 12.13). Young plants are adapted to the requirements of the ants (*Azteca mülleri*) by the queen biting through a thin point between the nodes — the prostoma, the pressure point of the axillary bud, which is shifted to a point opposite the bud by vertical growth — thus entering inside one of the chambers. The young workers of the developing colony later remove the partitions, thus extending their living space chamber by chamber. Apart from habitation, the Cecropi also offer food to the ants, in the form of 'Müller's corpuscles' (Müller,

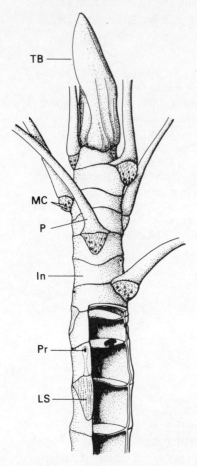

Fig. 12.13 Cecropia adenopus. In. internodum; LS. leaf scars; MC. Müller's corpuscles which lie in cushions of hair; P. petiole; Pr. prostoma; TB. terminal bud (after Gösswald, 1954 – from Eidmann, adapted).

1874, 1880; Belt, 1928; Schimper, 1898), structures rich in fats, protein and carbohydrates which grow at the base of the leaves, and which can be originally traced back to hydathodes, vegetable organisms which effectively excrete water in liquid form.

The advantages gained by the ants from the colonization of Cecropi are self-evident, but what of the Cecropi? Do they also gain an advantage from the guests? According to the investigations of Janzen (1969) into this question, Cecropi which are not inhabited by ants are usually overgrown with climbing plants and epiphytes, whilst those inhabited by ants are generally free of these. It emerges from Janzen's research that the ants are in fact responsible for this: he was able to show that the ants protect their nesting tree from overgrowing plants by chewing off the tops of all plants prevalent in that area, thus preventing them from growing any further. Apart from competing species of plants, herbivores may also be dangerous for the Cecropi, especially as the crowns of the young plant possess only 4-12 leaves. The investigations of Downhower (1975) into this indicate that under certain circumstances the leaves of plants with ants are less damaged by herbivores than those of *Cecropia* without ants.

These advantages enjoyed by *Cecropia* as a result of colonization by ants could represent a selective pressure in favour of an improvement of the living conditions of ants living in plants. In this case the advantage gained from the ants would be one of the decisive factors in the development of the 'Müller's corpuscles' as food source for the ants, for the formation of hollow chambers and the prostomae in the sprouting area between the nodes and for the lack of development of biochemical defence mechanisms against plant-sucking insects, the sweet liquid on which the ants live (Janzen, 1969).

In tropical Africa is the genus *Barteria,* which belongs to the Flacoutiaceae (Kohl, 1909; Bequaert, 1922). *Barteria fistulosa,* for example, is a tree with a maximum height of 15 m which lives in the shady areas of evergreen rain forests. Young plants of 1-1.5 m in height produce side branches initially of 10-20 cm in length. These side branches are hollow inside and seem slightly distended from the outside. They are sought out by young queens of the Pseudomyrmecine species *Pseudomyrmex aethiops* and *Pseudomyrmex latrifrons*: the queens gnaw holes in the soft branches and found their colonies inside. With around 20 workers, a colony is sufficiently strong to control the host plant and to successfully fight other young colonies which are not so far advanced in their development. There eventually remains only one colony on one branch, with a single queen remaining. The excreta of plant lice serve them as food, as well as fungi, both of which live in the hollow branches. As the young *Barteria* plant grows, the similarly growing colony migrates up to the crown of the tree. Fully developed colonies consist of 1000-4000 individuals.

In contrast to trees which had not been colonized, Janzen (1972) found that the inhabited trees had more leaves and branches, and that there was less destruction of leaves as a result of their being eaten. The plants in fact have the ants to thank for this unequivocal advantage, since, as the investigations of Janzen show, the ants not only remove all alien plants attempting to take over *Barteria* and drive away the widest variety of herbivores by means of their painful stings, but also scrupulously clean the leaves of whatever falls on them.

The best researched example of ant plants and their reciprocal effects with the ants inhabiting them are certain Acacia species, which are colonized exclusively by *Pseudomyrmex* species. The *Acacia* genus embraces about 300 species altogether, these being distributed throughout the tropical and subtropical regions of the world. Whilst the thorns of a number of African species swell greatly as a result of egg-laying butterflies, the larvae of which develop inside the swollen thorns, one group of *Acacia* species, found only in the New World, displays a spontaneous swelling of its hollow thorns (Fig. 12.14). Such

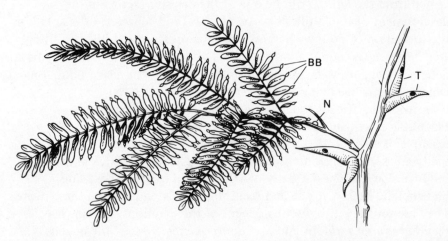

Fig. 12.14   Acacia sphaerocephala. The hollow thorns (T) of this species are mostly inhabited by ants, which gnaw entrance holes in the young thorns. BB. Belt's bodies; N. nectary (after Gösswald, 1954 — from Schimper, adapted).

swollen thorns are found, for example, among *Acacia cornigera, Acacia sphaerocephala, Acacia hindsii, Acacia collinsi* and *Acacia melanoceras* (Janzen, 1966). The thorns of these species are inhabited by specific *Pseudomyrmex* species, which are completely adapted to their host plants and never colonize any other species. Obligatory *Acacia* ants are *Pseudomyrmex ferruginea, Pseudomyrmex belti, Pseudomyrmex spinicola, Pseudomyrmex nigrocincta, Pseudomyrmex satanica* and *Pseudomyrmex nigropilosa* (Janzen, 1966). Apart from shelter, the ants also find their food on these plants, the food consisting mainly of 'Belt's bodies' (pear-like structures which grow at the tip of the pinnules) and the contents of nectaries, which are formed on the host plants outside the blooms (Fig. 12.14).

The relationship between *Acacia cornigera* and its inhabitant species *Pseudomyrmex ferruginea* was investigated by Janzen (1966). According to this research, colonization occurs when the plant is usually still very young, and is accomplished by inseminated *Pseudomyrmex* females, which gnaw holes in the still green thorns and then rear their first brood inside the thorns. During colony-founding, the queen leaves her nest continuously, gathering food for herself and the brood from the nectaries and also for the brood alone in the form of 'Belt's bodies'. As the colony grows in size, all the thorns are eventually occupied. After nine months, there have usually been so many workers produced (about 1200) that these are also obliged to patrol outside their host plants. After three years, the colony has grown to about 16 000 workers: a *Pseudomyrmex ferruginea* colony can include a maximum of 30 000 workers.

Apart from a few insect species which have developed mechanisms whereby they are able to eat in the presence of the ants, all insects on the *Acacia* are attacked by *Pseudomyrmex,* and usually successfully fought off. In addition to this, the ants attack all alien plants disturbing their host plant, chewing them with their mandibles. The same also occurs with plants growing within a radius of 10-150 cm from the plants inhabited by the ants. In this way a space is created for the plants inhabited by *Pseudomyrmex,* in which they are able to grow undisturbed. This results in these inhabited plants growing very quickly, far more quickly than other plants not inhabited by ants. Janzen was able to demonstrate that *Acacia cornigera,* in the course of its evolution, has lost the ability to withstand destruction by herbivores and pressure from competing plants. Plants of this species which are without ants are very feeble and grow so slowly that in the dense vegetation of their native environment they are not able to penetrate those areas with adequate sunlight. They are in fact able to bloom in these defective conditions, but are not able to produce seeds capable of development. There thus exists here a mutualism between ants and acacia plants. This mutualism evidently developed in the course of coevolution, which in the plant led to the formation of the characteristic thorns, Belt's bodies, leaf nectaries and year-round foliage — as a result of which the ants are provided with food throughout the year; the mutualism also resulted in aggressive behaviour in the ants towards alien vegetable and animal organisms, as well as consumption of the Belt's bodies, amongst other things.

Thus, whilst *Pseudomyrmex ferruginea* is completely monogynous, one finds in other *Pseudomyrmex* species, such as *Pseudomyrmex venefica,* a large number of queens. These polygnous *Pseudomyrmex* species have clearly developed from mongynous ancestors, the evolution from monogyny to polygyny signifying a substantial advantage for the acacia hosts (Janzen, 1973). In contrast to their monogynous related species, the females of the polygynous *Pseudomyrmex venefica* are inseminated by males flying past from other colonies, and on an older tree. The newly inseminated females then retire into the thorns from which they emerged, or else leave the tree and seek out unoccupied thorns on other plants. If these females encounter other colony-founding females of the same species on the plants, they fight each other much less than the females of monogynous species, but eventually join to form a joint colony. The result of this is that the young colony grows far more quickly than that of a monogynous species, and is thus effectively able to protect its host tree not after nine months, but as early as 3-5 months.

As a result of the large number of egg-laying females and new additional females joining the colony, it can grow to gigantic proportions, encompassing hundreds of trees, without aggressive behaviour arising between the inhabitants of even widely separated trees. If one treats all these inhabitants of the various neighbouring trees, all resulting from a single colony, then these are the largest ant colonies of which we know.

In contrast to *Pseudomyrmex ferruginea* and *Pseudomyrmex venefica,* there are other *Pseudomyrmex* species, e.g. *Pseudomyrmex nigropilosa,* which claim all the benefits of living in the acacia, but without fulfilling their 'duties' in return: they do not defend their acacia against other plants, or against herbivores. When animals of alien species approach them, they take refuge under leaves, or flee inside the thorns. Such inhabitants thus involve no protection for the acacia, which sooner or later die off. Until this happens, however, *Pseudomyrmex nigropilosa* produces sexually reproductive animals, which ensure the continuation of the species. Janzen (1975), who investigated this species, describes it as a parasite of mutualism.

12.5 Silk nests

One of the most interesting phenomena of ant biology is the

construction of silk nests. In contrast with the types of nest mentioned so far, in the construction of a silk nest the secretion of a silk gland, which solidifies quickly, is often used to fasten leaves together. Living leaves spun together in this way form the nest of *Oecophylla* (Fig. 12.15). Since, however, adult ants do not possess any silk glands, it was

Fig. 12.15   Oecophylla smaragdina nest, consisting of leaves woven together with the spinning secretion of the larvae (after Frisch, 1974, adapted).

not known initially where they obtained the silk for the construction of their nests. Doflein (1905) as well as Ridley (1890), Green (1896) and

Saville-Kent (1897) independently and before him, solved this riddle: the ants use their larvae as tools. Silk glands are not unusual in ant larvae; the larvae of many species have such glands which generally provide silk for the cocoon. This in fact no longer occurs with the larvae of many builders of silk nests (*Oecophylla* and *Polyrhachis;* Wheeler, 1915), but they nevertheless possess silk glands, which even as regards their extraordinary size surpass everything otherwise known of other ant larvae (Karawajew, 1929). These larvae are fetched by other workers and pressed on to the edges of the leaves, which are held together by other workers. They give out the secretion of their silk glands, serving simultaneously as looms and shuttles.

The ability to construct silk nests has been acquired by ants in at least three instances, independent of one another, these being *Oecophylla* (Hingston, 1927; Chauvin, 1952; Way, 1954a,b; Ledoux, 1949a,b, 1950; Mühlenberg and Maschwitz, 1973; Hemmingsen, 1973), *Polyrhachis* (Wasmann, 1905; Karawajew, 1928; Ledoux, 1958; Ofer, 1970) and *Camponotus* (Forel, 1905). All three genera belong to the Formicinae. It is possible that a fourth group of silk nest builders exists among the Dolichoderinae, perhaps *Technomyrmex bicolor textor* indigenous to Java. This was surmised by Jacobson and Forel (1909), but has still not been clarified beyond doubt.

Oecophylla nests exclusively in trees, living in gigantic colonies of well over 100 000 individuals. These gigantic colonies include a large number of nests (over 150), but contain only one queen, which is over 2 cm in length and grass green in colour. According to Ledoux (1949a,b, 1950), a colony may be formed in a variety of ways: by the independent founding of a colony by an inseminated female, by splitting off, where an inseminated female leaves the nest with a number of workers, or alternatively by a group of workers in a secondary nest taking a female which has flown in their direction, thus making themselves independent of the parent nest.

Nest building begins where workers have discovered a suitable place. They return to this place, seize with their mandibles the edges of neighbouring leaves and draw them close together. Larger distances between leaves are overcome by the workers taking other nest mates by the petiolus and forming chains in this manner (Fig. 12.16), which in the case of *Oecophylla longinoda* may consist of up to twelve ants (Way, 1954a). By means of these chains, which move slowly backwards, leaves situated greater distances apart are drawn together. Thus, while the leaves are held together by one group of workers — they can hold on to

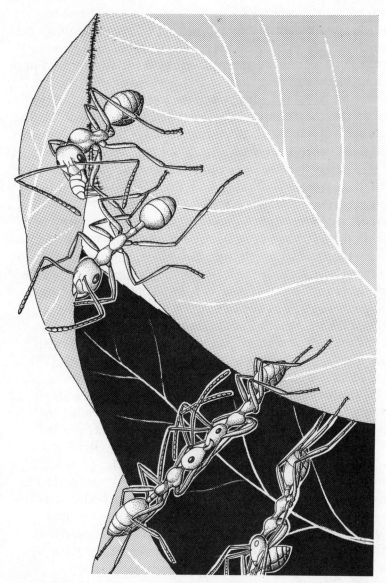

Fig. 12.16 Oecophylla workers constructing a nest (after Mühlenberg and Maschwitz, 1973, adapted).

these for over an hour – other workers fetch the larvae, control the edges of the leaves with their feelers and place the larvae alternately on the leaf edges which are to be joined together. With the emerging silk thread, leaves can be woven together, and gaps closed. This form of nest construction is chiefly a matter for the large workers among *Oecophylla*; the smaller ones are occupied with their tasks inside the nest, but are able to work at nest construction from inside (Hemmingsen, 1973). The finished nests are only inhabited as long as the leaves of which they are made remain fresh: when the leaves wilt and die, the ants leave the nest and move into a newly built one. *Oecophylla* even builds 'pavilions' for the plant lice, the pavilions containing no brood and constructed by means of spinning round closely situated leaves and twigs.

The Polyrhachis genus includes about 300 species, the majority of which build silk nests (Ofer, 1970), but only a small number of which have been more thoroughly investigated. *Polyrhachis gagates* nests in the ground in tropical Africa, using the web spun by the larvae to cover the walls and entrances to the nest (Ledoux, 1958). *Polyrhachis laboriosa,* on the other hand, a species similarly found in Africa, puts its nests, which are only a few centimetres in size, in trees, securing them with branch forks as a rule. The nests consist mainly of silk material, with bits of branch and twig, as well as leaves, spun into them (Fig. 12.17). New colonies are placed in the hollows of tree stumps and branches, into which the freshly inseminated females withdraw. They use their first larvae immediately to help with nest construction, covering the walls of their small dwellings with their secretions (Ledoux, 1958).

The nests of *Polyrhachis simplex* were investigated by Ofer (1970) in Israel. They are situated in various places, but mostly in gaps under stones, or in the hollows of trees; in the summer, *Polyrhachis* seeks the area around brooks, there putting up its summer quarters. The nests are hidden from the outside world by means of a silk curtain; a similar curtain separates the brood chamber from the waste heap inside the nest. In addition to accommodation for its own colonies, *Polyrhachis simplex* also constructs 'pavilions' of 2-5 cm in diameter, in which mostly Pseudococcides are kept. During actual spinning, the larvae are held by the workers at the rear end of their bodies, so that the front end is able to move freely. During this process, a fine silk thread is produced which attaches itself beneath, to previously spun webs and leaves drawn together by other workers. Further workers also stand by in the vicinity to press grains of sand and fine plant material into the web while it is still fresh.

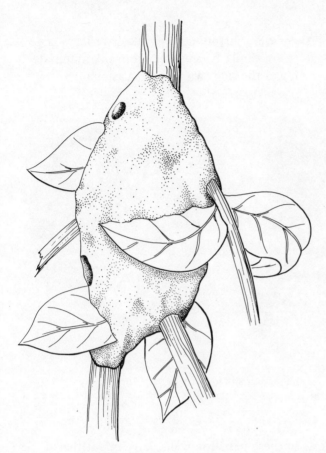

Fig. 12.17   Silk nest of *Polyrhachis laboriosa* (after Ledoux, 1958).

The silk nests of *Camponotus* have been even less well researched than those of *Polyrhachis* and *Oecophylla,* although Forel described the first, and so far only, indications of silk nests among *Camponotus* as early as 1905. On this occasion, Forel was sent a *Camponotus senex* nest from Brazil, the nest having been built into a tree and containing numerous chambers made of silk material. The sender of the nest (Goldi) also reported to Forel that he had observed the ants whilst weaving with their larvae.

Camponotus senex is not the only *Camponotus* species, however, which constructs spun nests. Thus about ten years ago the author received a *Camponotus (Myrmobrachys) formiciformis* FOR. nest, sent to him from the San Salvador region (Fig. 12.18). It was 11 cm high and had a diameter of 20 cm, and was built around the forked branch of a tree. The branches and twigs were surrounded by silk material; a large number of partition walls were similarly prepared from silk material, largely fixed to the branch and the side twigs, and occasionally

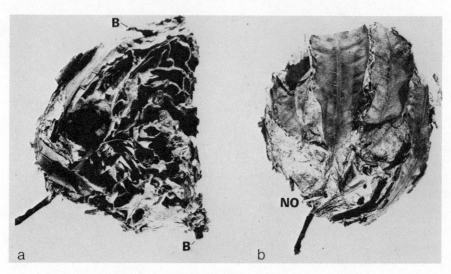

Fig. 12.18 Camponotus formicoformis silk nest. (a) section through the nest. B. branch around which the nest was constructed; NO. nest opening; (b) view on to the nest from the outside.

consisting of two layers. In these partition walls, large quantities of small pieces of leaf were to be found. The outer boundaries of the nest were formed from silk material, into which whole leaves had been spun. Openings in the outer walls had been parted for nest entrances.

13 **The nutrition of ants**

13.1 General summary

The food supply of ants is quite abundant: it consists mainly of animal and vegetable foods, which are taken by the various ant species in various ways and combinations. Their diet is usually composed of both animal and vegetable: the majority of species are neither wholly meat-eating, nor wholly vegetarian, and even those species which count as specialists for meat or vegetable food are not always consistent in this. This is demonstrated by using the African army ant species *Anomma molesta* as an example: this species brings back not only captured prey, but also eats fruits, such as bananas (Gotwald, 1974a). In addition to army ants, the Ponerines too were, for some time, considered to be exclusive flesh-eaters; however it was discovered that there were exceptions here also. We have since learned that *Odontomachus* also takes honeydew (Evans and Leston, 1971) and that *Rhytidoponera,* one of the Ponerines which is closely related to the Myrmicinae, also collects and utilizes seed kernels (Haskins, 1970).

Other insects including spiders, snails and worms usually serve the ants as a meat diet. Among the flesh-eating ants there are some which have specialized in a particular food source, such as the Cerapachynes, for example, which — as far as we know — live entirely on other ants (*see below*), or *Leptogenys elongata* which evidently captures only Isopods (crustaceans) as prey (Wheeler, 1904b), or *Leptogenys processionalis,* which has specialized in capturing termites (Wilson, after Maschwitz and Mühlenberg, 1973b).

Vegetable food is usually sucked up by the ants as plant juices, either directly from the plant, or indirectly via plant-sucking insects. When they take the plant juices directly, the ants reach them by means of the nectaries, which are formed in the plant, either inside or outside the blooms. In addition to this, many ants take in plant juices from open wounds in the plants, something which can frequently be observed in early spring. In the case of *Aphaenogaster rudis*, an unusual instance, for insects, of the use of intermediary means in order to obtain liquid food has been reported: the animals place small pieces of plant or stone on to the liquid food, and carry it into the nest later, with the liquid still attached (Fellers and Fellers, 1976).

As far as obtaining food indirectly from plant sucking insects – plant lice in particular – is concerned, the ants obtain the plant juices as a result of the fact that the plant lice excrete what for them are the unusable components, mostly sugar, in the form of honeydew. The plant lice give out this honeydew during mechanical contact with the ants: many ant species consume it, some of them in large quantities. An average *Formica polyctena* colony consumes about 155 l of honeydew per year, with a sugar content of 41 kg (Horstmann, 1974a). 'Trophobiontic relationships' of this kind, such as have formed between ants and plant lice, may also exist between ants and butterfly larvae, whereby the butterfly larvae release glandular secretions containing sugar (Malicky, 1969; Maschwitz *et al.,* 1975), and similarly between ants and plant-sucking Hemiptera (Maschwitz and Klinger, 1974). Food is taken more directly than using plant-sucking insects as intermediaries, by leaf-cutting ants, which cultivate a fungus, living mainly or exclusively on this, and also by harvester ants, which consume plant seeds for food.

The ingredients necessary for an adequate diet for ants were investigated by Bhatkar and Whitcomb (1970), by feeding ants with artificially mixed food. The most effective artificial food, with which over 30 different species from four different subfamilies were successfully fed, consisted of 5 g agar, 500 ml water, a whole hen's egg, 62 ml honey and one vitamin capsule (Mac Kesson Bexel). The vitamin capsule contained essentially the vitamins A, B1, B2, B6, B12, C, D and E and a number of mineral salts in addition. It is still not known what part these minerals and vitamins play in the nutrition of the ants: it was demonstrated, however, that the animals developed only slowly and incompletely without the vitamin capsule. Under the latter conditions, for example, *Formica schaufussi* produced no winged reproductives.

13.2 Ants as predators

Nearly all ants, with the exception of leaf-cutter and certain harvester species, live at least in part, by predation. Animal food is brought into the sometimes gigantic colonies of army ants in particularly large quantities. Wherever army ants are to be found, they count among the greatest predators of their biotope. According to Topoff (1972), a single *Eciton burchelli* colony consumes over 100 000 captured animals in a day, these consisting mainly of other insects.

The predatory expeditions of *Eciton burchelli* begin at dawn during the nomadic phase. At this time, the workers leave their bivouac and divide themselves into a number of processions, each of which in its turn divides up into a network of individual branches. These processions are mostly formed from medium-sized workers, the smaller ones remaining in the bivouac. The largest workers, the soldiers, patrol the length of the predatory processions. They do not carry any prey, serving, with their powerful mandibles, exclusively for the protection of the colony (*see* Fig. 2.3). Anything which can be seized in the way of insects and other predated animals is attacked by the workers at the head of the army. The prey is killed by means of bites and stings, is torn into pieces and transported back to the bivouac. A traffic in the opposite direction is thus produced in the predatory army of the ants: whilst one section of the workers presses forward, others return to the bivouac and give the prey to smaller workers, which then feed it to the larvae (Topoff, 1972).

If one disregards the differences between the various groups and species of army ant, then there are essentially two features characteristic of the life habits of the Dorylines, these being nomadism and group predation. In both cases, there are preliminary levels and expressions in other subfamilies, which give indications of the development of the Dorylines and the emergence of their lifestyle (Wilson, 1958a).

The term nomadism is used when a colony frequently changes its nesting place: this behaviour is found in other subfamilies, but not with such frequency as with the Dorylines. The very primitive Ponerine species *Amblyopone pallipes,* which lives mainly on centipedes, displays an interesting type of nomadism. Workers of this species were often found in groups around freshly killed prey, from which their larvae were eating. Closer examination of these groups revealed that here it was not the prey which was brought to the larvae, but the larvae which were transported to the freshly killed prey. Similar behaviour was found by Wilson (1958a) among *Myopone castanea,* a species also belonging to the Amblyoponini, in New Guinea.

Under the single term 'group predation', Wilson (1958a) is in fact putting two different things together, these being the joint attack of prey animals and the join transportation of the captured prey back into the nest. Whilst the joint transportation of prey occurs relatively frequently among the more highly evolved species, there are in fact only a few Ponerines which also attack in groups. Those that do are Ponerines which capture termites, such as a number of species from the *Leptogenys, Megaponera, Paltothyreus* and *Ophthalmopone* genera. Cerapachynes, which exclusively capture other ants, also belong to this group of ants which hunt in groups.

It is interesting, therefore, that there are also species outside the Dorylines, in the Ponerine subfamily, which both live nomadically and engage in group predation, thus behaving, in certain essential features, like army ants. One example of this behaviour is *Leptogenys diminuta,* observed by Wilson (1958a) in New Guinea. This species does not build any nests, inhabiting already-existing holes in the ground, or simply living amongst twigs and foliage. How long they remain in one place is not yet known, since no nest changes have so far been observed, either by Wilson or by Mann, who had done research into this species before Wilson in 1919. We do know from the observations of Wilson, however, that *Leptogenys diminuta* hunts in groups. The groups consist of about 40 workers, hunting in close formation in lines of 3-6 ants. After a successful hunt, the captured prey is brought back to the nesting place by the group. These observations by Wilson are supported by those of Maschwitz and Mühlenberg (1975), who observed predatory expeditions of 3-100 animals. They were able to demonstrate that these predatory groups follow trails laid by scouts between the prey and the nest. Occasionally, however, the columns go out in vain, if the prey has disappeared from the place to which the trails originally led.

Leptogenys ocellifera operates more economically, by putting down lasting trails in the vicinity of its nests, travelled day and night by workers and watched by other sentry workers. If workers hunting individually or in groups then capture prey, they do not need to bring it back to the nest but only as far as the lasting trail in order to obtain reinforcements. On their path between the prey and the lasting trail, the workers lay a trail with the secretion of their poison gland, which functions simultaneously for orientation and raising the alarm. In this way, they attract other workers from the lasting trail to the prey, so effectively in fact that even a few minutes after the prey has been discovered, dozens, sometimes even hundreds of workers find their way to it.

Hunting in groups, ants are often able to capture larger animals, or the brood of other social insects, which for ants hunting individually would be unobtainable (Wilson, 1958a). *Leptogenys diminuta* and *Leptogenys purpurea* have specialized in the capture of the relatively larger Arthropods: *Megaponera* in the capture of termites, the Cerapachynes in the capture of ants and the Doryline *Eciton* and *Anomma* groups in a whole range of other Arthropods, including social wasps and ants.

The connection between group predation and nomadism could — according to Wilson, 1958a — have its roots in the fact that larger Arthropods and social insects are distributed over a wider area than smaller prey. The ants hunting in groups must therefore correspondingly extend their hunting areas, which could have led to nomadism. This hypothesis, however, is not appropriate in the case of the *Leptogenys* genus, since among *Leptogenys binghami* and *Leptogenys attenuata,* for example, nest change-overs are very well organized, but a typical group hunting behaviour has not formed. In addition, *Leptogenys ocellifera* does not change its nest sites more frequently, in spite of its highly developed hunting methods, than *Leptogenys binghami,* which hunts individually and moreover captures social insects (termites) (Maschwitz and Mühlenberg, 1975)

The mandibles of certain ant species are characteristically formed by way of adaptation to their lifestyle: there are thus species whose mandibles open almost to an angle of 180° (Fig. 13.1) and which can close vigorously by means of powerful closing muscles (Fig. 13.2), whilst the teeth of the mandibles sink deep into the prey. Mandibles such as these have formed in several ant genera in various subfamilies, and certainly appeared independently of each other. They are found in *Odontomachus* and *Anochaetus* (Ponerine), in the *Dacetini* (Myrmicines) and in *Myrmoteras* (Formicines).

The *Dacetini* consist of about 200 known species, which either mostly or exclusively capture Springtails (Collembola) as prey (Brown and Wilson, 1959). The Collembola belong to the primary unwinged group of the primitive insects, most of which are equipped with a tail-spring: this is folded forwards under the abodmen in the resting position. When the animals are disturbed, the tail-spring is suddenly stuck behind, and they leap away. These Collembola are captured by the *Dacetini* either in the ground itself, or on the surface. Since the animals spring away at even a small disturbance, the ants creep up to them very cautiously, their mandibles spread apart. The mandibles in the case of *Strumigenys* and other species are opened by means of a tooth at the basis of the mandible, which rests in a lateral protruberance

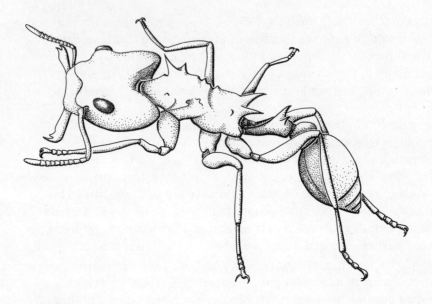

Fig. 13.1   *Daceton armigerum* worker with parted mandibles (after Wilson, 1962a, adapted).

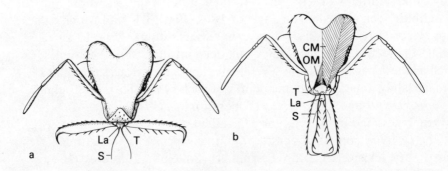

Fig. 13.2   Head of a *Strumigenys loriae* worker with opened (a) and closed (b) mandibles. CM. closing muscle of the mandible; La. labrum; OM. opening muscle; S. sensory bristles; T. tooth-like appendix of the mandible which rests behind a lateral projection of the labrum when the mandibles are open (after Brown and Wilson, 1959, adapted).

of the hard labrum (Fig. 13.2). By means of this, the mandibles are opened wide, remaining at the same time under tension from the powerful closing muscles. In the centre of the labrum are two long, sensory bristles, which are only disturbed by the Collembola when they are already in the grasping range of the mandibles. The disturbance of the bristles leads to a rhythmic opening and shutting of the mandibles, the Collembola being unable to release themselves by leaping away. This beating with mandibles and their teeth is frequently sufficient to render the prey incapable of movement. If this is not the case, the still-struggling prey is carried up from under the ground and killed with a sting (Brown and Wilson, 1959).

Daceton armigerum (Fig. 13.1) hunts in trees and shrubs, so the mandibles of this species are usually opened to an angle of only 45-90°. In the case of greater disturbance, however, they are opened wider: they move outwards and are held open by the upper lip, thus now forming an angle of 170°. Faced with larger objects, which cannot be considered as prey by the ants, they flee, displaying on these occasions, their unusual ability to run as quickly backwards as forwards. Objects of prey are first felt quickly with the antennae and then seized by the sudden closing of the mandibles. Because of particularly long claws on their feet, the *Daceton* workers have a good grip on the trees and shrubs, and are thus also accomplished at fighting larger prey animals, even when these defend themselves fiercely (Wilson, 1962a).

Ants whose mandibles are not specialized in this manner overcome their prey in a different way. We already know of the Formicine weaver ant *Oecophylla longinoda,* from the researches of Cohic (1948), Weber (1949) and Vanderplank (1960), that in addition to other insects, they also capture army ants of the *Anomma* genus. This does not occur by means of a joint effort, but is the work of individual workers, who sit on twigs or leaves beside the army ants as they pass by. They then suddenly attack the mass of *Anomma* workers with their mandibles, dragging one of them out. The extremities of the attacked army ant are then torn at by several *Oecophylla* workers − probably brought to the scene by means of alarm signals − until the ant has been immobilized. It is then carried back to the *Oecophylla* nest.

This behavioural trait, stretching the victim to death with the combined strength of the group, is also displayed by other ant species. According to Schneirla (1971), this behaviour occurs among the army ant species *Eciton burchelli* and *Labidus coecus,* which not only kill larger prey animals in this way, but also actually tear them into pieces. Among wood ants and other species (Fig. 13.3), one may frequently observe how prey animals, and intruders into the colony, are attacked and stretched at the extremities.

Fig. 13.3   Leptothorax workers attacking an ant from an alien colony (*Harpagoxenus*) by stretching its feelers and legs (Photo — A. Buschinger).

13.3 Trophobiotic connections between ants and plant lice

Honeydew, the excrement of plant lice, plays an extraordinarily large part in the diet of many ants, especially in the Dolichoderine, Myrmicine and Formicine subfamilies. Almost all the central European species are so dependent on the honeydew in their diet that they only inhabit areas where there are plant lice. In the trophobiotic relationship between aphids and ants, as it is understood by Kloft (1959a), the excrement is extracted by the ants directly from the anal region of the lice (Fig. 13.4).

For the aphids themselves, the giving of their excrement presents particular problems: they have an enormous flow of food through their bodies, taking in a great deal of liquid and thus also excreting a great deal. Since the excrement has a high sugar content, the aphids producing it, which frequently live in colonies, may easily become sticky with the fluid, or make others so. They have consequently developed a variety of mechanisms for ridding themselves of the excrement without mishap: a number of species spin it into wax threads, others spray it in long jets from the anus, whilst still others use their hind legs as an aid in hurling it away (Kunkel, 1972).

Many aphids have adapted to the ants which take their excrement during the course of living with them. The wax glands have thus reduced

in a number of species whose antecedents had spun wax threads: since
the ants provided a harmless means of elimination, the glands had thus
become superfluous. In other species, there has been increased growth
of the hair around the anus, which is an advantage for the ants taking

Fig. 13.4   Wood ant worker, 'milking' a greenfly (after von Frisch, 1974, adapted).

the excrement directly from the body of the aphids (Schmidt, 1952).
In the course of living with the ants, the so-called cauda (Fig. 13.5) has
also been reduced, this probably having previously protected the back
of the aphid from rolling excrement (Kunkel, 1973).

The behavioural alterations brought about in the aphids as a result of
living with ants first concern their readiness to give out the excrement:

they withhold it, but at the same time indicate their readiness to give
the honeydew by means of specific signals, by raising the rear section of
their bodies and swinging their legs. Other plant lice have reduced these

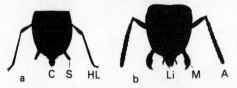

Fig. 13.5   General comparison between the abdomen of a greenfly (a), seen from
the ant's perspective, and the head of an ant offering food (b). A. antennae;
C. cauda; HL. hind leg; Li. labium; M. mandible; S. siphon (after Kloft, 1959a).

signals to the minimum inconvenience. The signals from the ants which
induce the lice to give the excrement are very specialized in many cases.
Among these species, a signal from the ant by means of imitating the
mechanical stimulation of a fine bristle is not sufficient to produce the
excrement, whilst among other aphid species, this is enough, as is the
case with many Lachnids. These give out a drop of excrement at a
gentle mechanical stimulation, but immediately withdraw it if it is not
taken. Many root lice, which have associated particularly closely with
ants, also give out a drop of excrement at a mechanical signal (Kunkel,
1973).

Among some species, living with ants has led to the aphids altering
their life rhythms: we know of anholocyclic plant lice which only
reproduce parthenogenetically, and holocyclic aphids which reproduce
alternately parthenogenetically and sexually. These differing generations
may remain on the same plant (monoezic species), or they may change
their host plant (heteroezic species). One heteroezic species, for
example, is *Aphis fabae,* which is not associated with ants. *Disaphis
bonomii,* on the other hand, has become secondarily monoezic in the
course of its adaptation to trophobiosis, and is thus at the disposal of
the ants for several generations (Way, 1963).

The birch louse *Symydobius oblongus* only lives with ants: if the ants
are kept away from the aphids, then they die out after a few days.
Aphids which are not visited by ants initially hold back their excrement;
they become restless, their reactions varying among individuals. Some of
them raise the rear section of their bodies perpendicular to the ground,
which they do not usually do as long as they are being visited by ants.
The larvae soon begin to swing their hind legs back and forth. Some

aphids display a rhythmic movement of the anal flap, accompanied by a drop of excrement moving in and out of the orifice. Eventually they rid themselves of the excrement, frequently however neither flinging it with their feet nor spraying it away, so that a number of individual aphids are soon stuck to the birch tree. Among some populations, young larvae may, however, begin to fling away the excrement, but in general restlessness in the colony and attempts to leave increase.

The signals used by the plant lice offering excrement to the ants have been compared by Kloft (1959b, 1960) with the signals of ants offering food during the exchange of food. As has already been described, the ants offer their food with widely opened mandibles, outstretched lower lip and light touches of their feelers. It is thus striking that similar signals are noticeable when aphids are offering excrement, as is shown in Fig. 13.5. This illustration outlines the rear body section of a plant louse, and the head of an ant offering food. The similarities between these two sketches lie in the fact that the hind legs of the aphid resemble the ant feelers, the two siphons resemble the parted mandibles and the cauda the outstretched lower lip of the ant. In addition to these visual similarities, the aphids frequently swing their hind legs, as the ants offering food do with their feelers. The reaction of the ants to the plant louse ready to give its excrement similarly suits the trophallaxis schema: they display begging behaviour, as they would towards ants offering food, thus inducing the aphids to give the drop of excrement. Against this hypothesis of Kloft, that aphids imitate the stimulus schema of ants offering food when they are offering excrement, is the fact that there are also trophobiotic connections with Coccines, which do not fit into this schema at all.

The advantages for the aphids which have grown out of their trophobiosis with ants do not merely rest in the harmless evacuation of their excrement: the ants also protect their aphids from their enemies, as is shown in the research of El-Ziady and Kennedy (1956). It was revealed that *Lasius niger* workers attacked and removed the larvae of Coccinellids (ladybirds) and Syrphids, which live on aphids, when they were in the vicinity of their own aphids. Similar results have emerged for numerous other ant species. The pavilions which are built by species such as *Oecophylla* for the protection of their coccids are not only a protection against the natural enemies of the plant lice, but also against unfavourable atmospheric conditions. This emerged from the investigations of Way (1954b), according to which *Oecophylla longinoda* builds its 'louse stalls' most often during the wettest time of the year and when there is rainfall. The stalls are not in fact, waterproof, but they do nevertheless prevent the aphids from being washed away by the rain. We know from certain species of the harvester ant genus *Pheidole* that these protective structures are not only put up by the ants over the aphids themselves, but also over ant paths leading to the aphids. When there is strong wind or rain, only protected plant lice like these are sought by the ants. This would seem to indicate that these protective structures are primarily for the protection of the ants' own colony, and only secondarily for the protection of the aphids (Way, 1963).

Among some aphid species which are kept by ants, the ants collect the eggs from the host plants and bring them into their nests for the duration of the winter (Zwölfer, 1958; Pontin, 1960b). A number of aphid species are only able to survive the winter in this way. The lice which hatch from the eggs are brought into the brood chambers in the spring and find their way from here back to the host plants by themselves.

A remarkable relationship between certain plant lice of the *Hippeococcus* genus and ants of the *Dolichoderus* genus in Java was reported by Reyne (1954). All known species of this genus are associated with ants, which seek out their plant lice on branches and bark. When the plant lice are disturbed, the larvae climb on to the backs of the ants, whilst other lice are grasped between the mandibles of the ants and taken away.

In general, the relationship between ants and aphids displays certain features which show parallels with the rules of domestication (Kunkel, 1973):

Both partners are social to a certain degree, and are able to communicate with each other: the movement of the hind legs by the aphids fits into the schema of the feeler movement of ants offering food.

The relationship is only changed to the extent of being modified in the course of domestication: this is shown in the giving of excrement by the aphids, in which, when ants are not present, some previous behavioural forms return.

Under the influence of the ties between them, the partners are withdrawn from their natural influence: this gives rise to certain mutants which form a dependent relationship, as has been demonstrated for aphids.

In a domestication relationship, mutations of neutral value may also be obtained, which leads to greater variability: this may also be observed among certain aphid species, for example Lachnids.

What, however, do the ants gain from their connection with the plant lice? Of what does the honeydew consist, it being essential for the nutrition of many ants? Honeydew consists essentially of sugar, in fact mainly fructose, sucrose and glucose, some maltose and various chains of these constituents (oligosaccharides). In addition, there are also free amino acids, proteins, minerals and B vitamins. As far as the amino acids are concerned, they may amount to over 13% of the dry weight of the honeydew and include all the amino acids necessary for animal nutrition. Differential combinations of ingredients in the honeydew are based less on the different species of aphid connected with the ants, and more on the different species of host plants, their age and differences in parts of the plants which are tapped by the aphids (summary: Way, 1963; Auclair, 1963).

13.4 Harvester ants

Harvester ants have been observed for a long time looking for cereals and other plant seeds and storing them in their nests, sometimes in large quantities. This kind of storage system, as practised by harvester ants, is only typical for certain species, in particular those which have to survive long periods of drought. Other species do not have to do this in the areas they inhabit, either because there is sufficient food available to them all year round, or because the time of year when there is little food coincides with the cold season, which is endured by the ants by means of a cold season rigour.

The collection of plant seeds in the nest, however, is not necessarily linked with a storage regime. This can be seen from the many examples of central European species, such as *Tetramorium caespitum, Lasius fuliginosus, Lasius niger, Formica fusca, Formica exsecta,* many wood ant species and a number of *Myrmica* species (Wellenstein, 1952), which occasionally collect plant seeds. In these cases, however, the plant seeds are not 'put in the cellar', but are used immediately. Also, in contrast to the typical harvester ants, these species are not interested in the whole seed, only in parts of them, supplementary oil content, which are called elaiosomes.

The plant seeds are more important for the harvester species of the genera *Messor* Fig. 13.6) and *Monomorium,* indigenous to the Mediterranean area, for the American harvester ants of the *Pogonomyrmex* genus and for the *Veromessor* genus, similarly found in

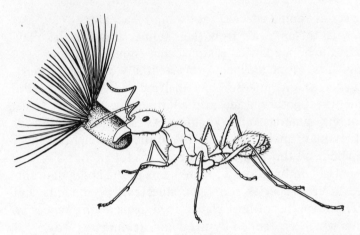

Fig. 13.6　*Messor* worker bringing back seeds (after Linsenmaier, 1972, adapted).

America. Another genus of harvester ants (*Pheidole*) has species
distributed both throughout the New and the Old World. For these ants
also, plant seeds do not represent the only food source; they sometimes
also bring back prey, and will take sugar water if it is offered to them.
The plant seeds which they store, however, are necessary for their
nutrition during the drought which these species normally have to
survive. By chewing the contents of the seeds and by salivating it with
the enzyme rich secretion of their labial glands, they successfully
convert the starch stored in the seeds into sugar. The sugar obtained in
this way essentially represents that which in other species would be
obtained from aphids in the form of honeydew.

The preparation of the seeds in the nest usually begins with the seeds
being opened and the shells being removed. The contents of the seeds
are then chewed in a joint effort and mixed with the secretion of the
labial glands. By means of this process, the starches are converted first
into maltose, and then glucose (Delage, 1962). The germination of the
seeds, which are stored by the ants in their warm, moist nests, is
probably prevented by *Messor* by means of the secretion of their
metathoracal glands, which, according to research conducted by Koob
(1971) and Schildknecht and Koob (1970, 1971) contains, amongst
other things, β-hydroxydecane acids. This substance, called myrmicacin,
has a strong inhibiting effect on germination, as one can obviously see
from a 'plum pulp test': if one treats plum pulp with myrmicacin, then,
even after 20 days, it will not have been attacked by fungi, or fruit flies,
whilst untreated pulp is completely ruined after this time. With this

germination-inhibiting myrmicacin, therefore, the ants are able to prevent the germination of the seeds they bring back to the nest.

13.5 Leaf-cutter ants

In the same way as the obligatory symbiosis of many ant species with certain species of aphid and the symbiosis of many obligatory plant-inhabiting species with their host plants, so also the relationship of leaf-cutter ants with the fungi on which they feed is an obligatory mutualism between very different organisms (Martin, 1969). Leaf-cutter ants live exclusively in the tropical and sub-tropical regions of the New World. All species belonging to this group grow their fungi in their nests, living mainly, if not exclusively, on what they grow. The fungi are cultivated on a culture substratum, the nature of which differs widely amongst different species: the simpler and more primitive species use insect faeces or even dead insects as a substratum, whilst more highly developed species cultivate their fungi on pieces of dead plant or on leaves and pieces of flowers which have been cut off the living plant. Species of the *Acromyrmex* and *Atta* genera are the most highly developed: they use almost exclusively leaves or parts of flowers from living plants as the substratum for their fungi culture.

The origins of fungus cultivation among ants have still not been completely explained. According to the hypothesis of Ihering (1894), leaf-cutter ants can be traced back to the harvester ants. According to this representation, the unavoidable mould growth on the still unripe seeds brought back to the nest was the main departure point for fungus cultivation. A proportion of the harvester species preferred this fungal growth, and eventually changed to cultivating the fungi on another substratum. Forel (1902), on the other hand, wrote that the leaf-cutter ant can be traced back to species which originally colonized in rotting wood and fed on the fungi which grew there. It is more probable, however, that the fungus cultivation of the leaf-cutter ant stemmed from those species which changed over to feeding on the fungi which grew on their own faeces (Weber, 1956b). This level of development is represented by species of *Allumerus,* which colonize in myrmecodomati in *Hirtella* (Rosaceae): *Allumerus* species cultivate a fungus on their own faeces and nest refuse, during which the particularly long and strong bristles of the host plants serve as 'bean poles' (Vogel, pers. inf.).

Although the symbiosis of the leaf-cutter ants with their fungi has been known since the time of the investigations of Möller (1893), we are still unclear about the identity of the species of fungi cultivated by the ants. Möller (1893), Wheeler (1907) and Weber (1972a) thus assume that the fungi of the leaf-cutter ants belong to the higher fungi (basidiomycetes), whilst Lehmann (1974, 1975) takes them to be ascomycetes, placing them in the *Aspergillus flavus* group. In fact the indentification of these fungi presents particular difficulties, since they do not produce any spore carriers in the ant nests, and because their cultivation on artificial culture mediums has proved particularly difficult. On the customary culture medium (potato-dextrose or dextrose-agar), the fungus grows only very slowly, and soon becomes overgrown by other micro-organisms. As soon as the ants have access to their fungi, however, the culture survives for an unlimited time.

In the nests of leaf-cutter ants, for example, *Atta sexdens,* the fungi are cultivated in a large number of chambers. The individual cultures have a brittle, spongy structure and are usually semi-circular, with a diameter of 15-30 cm. The individual mycelia are continuously being cut by the ants, thus forming nodular proliferations, which are called 'ambrosia bodies' or 'kohlrabi heads' (Möller, 1893). It is these tiny white structures which serve the ants as food (Fig. 13.7). The vegetable

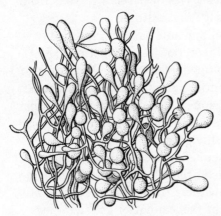

Fig. 13.7 'Ambrosia' bodies of a leaf-cutter fungus (after Wheeler, 1923, adapted).

material brought in by a species such as *Atta cephalotes* consists of 49% freshly cut leaves, 44% fresh blooms, 3% cut foliage and less than 1% twigs; the remaining 3% of the material has not been identified (Cherrett, 1972). An important factor in the cutting and bringing-in of the vegetable material — apart from its physical condition — is the presence in it of specific ingredients: if substances such as sucrose, rhamnose, trehalose and arabinose are dropped on to filter paper, then this too will be cut up and brought in by the ants (Littledyke and Cherrett, 1975). During cutting itself, pieces are cut out of leaves and flowers and carried upright between the mandibles back to the nest (Fig. 13.8). Inside the nest, the pieces are put together, cut into small

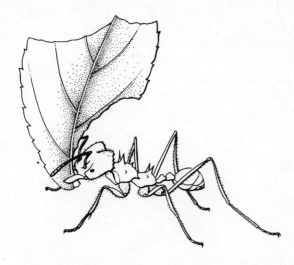

Fig. 13.8   Leaf-cutter ant carrying back a piece of leaf (after Linsenmaier, 1972, adapted).

pieces and salivated. Before they are placed in the fungus garden and covered with mycelia, the ants place them at the tip of their rear segment and excrete a drop of faecal material on to them. In this way, they evidently carry substances on to the substratum which are necessary for the growth of fungi (Weber, 1947, 1956a,b, 1958, 1966).

Weber (1947, 1966) at first assumed that the saliva glands or the faeces of the leaf-cutter ants contained antibiotics which prevented the growth of alien micro-organisms in the fungus garden: it was not possible, however, to confirm the validity of this hypothesis in the investigations of Martin *et al.* (1969). No indications of the presence of antibiotics could be found, either in extracts of the species *Atta colombica tonsipes* which he investigated, or in the remains of old cultures which had been thrown out of the nest, but they were found in the metathoracal glands. The reservoirs for these glands (Fig. 13.9), into which several other glands open, have no closure, so that their

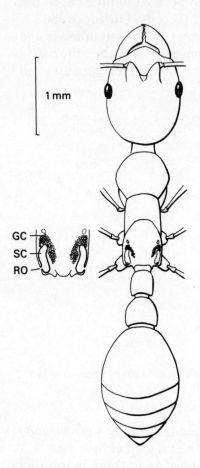

Fig. 13.9   Metathoracal gland of *Myrmica. Right* − general view, *left* − section of
the metathorax. GC. glandular cells; RO. reservoir opening; SC. secretion cavity
(from Maschwitz, unpublished material, after Janet, 1898).

secretion is distributed over the body surface of the ants: the secretion
of the metathoracal gland proved to be acid, with strong antibiotic
properties (Maschwitz *et al.*, 1970; Maschwitz, 1971). In leaf-cutter
ants, the secretion contains phenylacetic acid, which is even more
effective than benzoic acid. Apart from these, the secretion contains
further substances with antibiotic properties, such as β-hydroxydecane
acid (myrmicacin), β-hydroxyoctane acid and β-hydroxyhexane acid,

and in addition the plant growth substance β-indolyacetic acid (Koob, 1971).

Of what, however, does the substance in the faeces of the leaf-cutter ants consist, which evidently is also necessary for the growth of the fungi? According to the research of Martin and Martin (1970), mainly allantoines are found in the faeces, as well as ammonium, all the free amino acids found in the natural environment and a number of enzymes, such as α-amylase, chitinase and several proteinases (Martin and Martin, 1970, 1971; Martin *et al.*, 1973).

It was first thought that the ants compensated for the absence of these substances in the fungi themselves with these enzymes which are present in their faeces, in particular the proteolytic enzymes, thus facilitating their growth in the cultures. However, according to investigations conducted by Boyd and Martin (1975b) and by Martin *et al.* (1975), the same proteolytic enzymes proven to exist in the ant faeces are, in fact, also found in the mycelia of the fungi. Therefore we can conclude that the ants drew the enzymes indicated from the fungi, and not the fungi from the ants. This conclusion is supported by the results of Boyd and Martin (1975b) and of Martin *et al.* (1975), according to which the combination of the proteolytic enzymes remains constant throughout the entire digestive tract of the ants, thus passing through unchanged.

The ant-fungus symbiosis does not rest on the fact that the ants tend their fungi with enzymes, but on the fact that they transport the enzymes to those places in the fungus garden where they are most urgently required, that is where extensions are being constructed in the garden and the fungal mycelia being placed on freshly laid substratum. The ants take up the enzymes with their food from the quickly growing parts of the fungal garden, thus fertilizing the new substratum with their excreta, so that the freshly planted fungal mycelia take up the substances necessary for their growth more quickly and easily, thus growing rapidly right from the start. A proportion of the enzymes taken away by the ants are held back when possible in order to compensate for the absence of enzymes in their own bodies.

According to Garrett (1950; quoted by Schmid, 1971), there are five main factors which contribute to the ability of fungus to survive: 1. a strong inoculation potential, which enables the fungus to quickly overcome substratum resistance; 2. quick hyphen growth, which is favourable to the speedy colonization and extension over the substratum; 3. strong enzyme production, which facilitates the speedy utilization of foodstuffs; 4. antibiotic production to reduce competition; and 5. high tolerance of toxins produced by other organisms. The contribution of the ants to the vigorous growth of their fungi consists first in the fact that they procure vegetable material, prepare it, by removing and destroying alien micro-organisms, enrich the material with foodstuffs for the initial growth of the fungus, plant the fungus itself in the vegetable material, procure enzymes from other parts of the fungus garden, where the fungus is already in a fast stage of growth, and finally, remove competing organisms and used substratum from the culture. In return, the ants obtain not only food from the fungi, but probably also enzymes which are lacking in their own bodies.

By growing a fungus which decomposes cellulose, the leaf-cutter ants are able to make use of the great cellulose reserves of the rain forests (Martin and Weber, 1969). What termites manage to achieve by having endosymbiotic microbes which break down cellulose, the leaf-cutter ants achieve by means of their Ectosymbiotes, through their mutualistic relationship with fungi which break down cellulose.

14 Weapons and defensive behaviour of ants

Ants have at their disposal a variety of weapons and defence mechanisms which they bring into play when fighting prey and each other, as well as when defending themselves against predators. The large number of these defence mechanisms, far greater than amongst other animal groups, is probably — according to Maschwitz, 1975a — linked to the fact that the social insects not only have themselves to defend, but also the other members of their society. The weapons as such are partly those which are purely mechanical, partly mechanical weapons in conjunction with chemical weapons and, finally, purely chemical weapons.

Belonging to the group of purely mechanical weapons are mandrels on the thorax and the petiolus, which are widely distributed among ants, being particularly noticeably expressed among *Polyrhachis*. The mandrels are sometimes extremely sharp and can make themselves felt when one takes hold of the animal. As a result of the mandrels, the ants became completely proficient at self-defence, being captured as prey only very seldom, and when they are, spat out again by their predators. In some species, such as *Polyrhachis lamellidens, Polyrhachis furcata, Polyrhachis ypsilon, Polyrhachis bihamata* and *Polyrhachis craddocki* (Fig. 14.1), the mandrels, particularly those on the petiolus, are bent over to form hooks, which may dig into the skin of the attacking predators. This does not, in fact, give any advantage to the ant being attacked, but it may prevent the predators from attacking further animals from the same colony, thus serving the colony itself and the total population (autothysia; *see below*). The effect of the mandrels on attacking predators, however, has not yet been investigated in detail.

The sting possessed by ants can be traced back originally to ovipositor apparatus which is still functional among species distantly related to ants, belonging to the *Symphyta*. This apparatus has a very complex structure, which we do not yet fully understand (*see* summary by Oeser, 1961; Maschwitz and Kloft, 1971). The extendable sting is enclosed in a sheath, and is connected to the poison and Dufour's

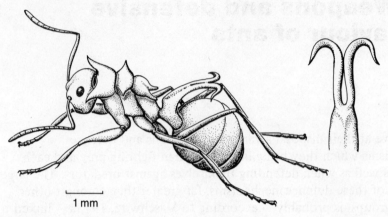

1 mm

Fig. 14.1   Polyrhachis craddocki worker and enlarged representation of the petiolus hook (after Bingham, 1903, adapted).

glands, which release their secretions in small amounts through a lamina, which serves simultaneously as a pumping membrane and a vent. Secretions composed mostly of toxic proteins are injected into the opponent at this time (Cavill and Robertson, 1965).

The stinging apparatus of ants is nevertheless reduced in the most highly developed groups of species, the chitin elements in these cases being generally more developed, whilst the glandular portions are reformed very differently. Only the Ponerines, the Cerapachyines, the Myrmeciines, the Pseudomyrmecines and the Aneuretines still have a functioning sting, whilst among the Dorylines, essentially only the South American Ecitons are able to sting: all the Dolichoderines and Formicines, on the other hand, are no longer able to sting (Foerster, 1912; Wilson *et al.* 1956; Stumper, 1960; Blum, 1966). The stinging capability of *Aneuretus* is interesting, since this species represents a stage of development between the non-stinging Dolichoderines and the Formicines. It demonstrates that sting reduction among ants frequently occurred independently among the species. Among the Myrmicines, it is mainly the leaf-cutter ants and some of the *Aphaeonogaster, Pheidole* and *Crematogaster* species which have lost their ability to sting.

The question as to why a weapon should become reduced amongst ants which has clearly proved its worth so well amonst bees and wasps was more closely investigated by Maschwitz (unpublished). The most immediately obvious assumption, that the sting reduction is connected with a change in diet from animal to vegetable food, was definitely proved incorrect: had this been the case, it would not have been possible for the sting to be reduced in a large proportion of the Dorylines, which live almost exclusively on predated animals. Sting reduction does not, in all probability, have much to do with defence against ant predators, since if this had been the case, one would expect to find numerous differences between the stomachs of insect-eating mammals which ate stinging ants and those which did not, but no such differences could be proved to exist.

If, however, one compares stinging and non-stinging ants when fighting each other, it becomes clear that the non-stinging ants have a distinct advantage over their stinging opponents: the animals which could sting were only able to pierce the smooth, mobile surface of the ants' bodies with great difficulty, according to the observations of Maschwitz (unpublished), and it was sometimes over half an hour before they could bring their sting to the segment edge of their opponents. It also emerges from the research of Bhatkar *et al.* (1972) that the ant stings are not an optimally effective weapon against arthropods: the reduction of the sting can probably be explained by the inadequacy of the stinging apparatus against other arthropods, and in particular against other competing ant species, as well as by the parallel development of new fighting methods at the same time as the sting itself was reducing.

The new fighting methods which give the non-stinging ants an advantage in combat with stinging species are fighting with the mandibles and, in addition, new chemical weapons. There are transitional stages between those two new types, these being sometimes adopted simultaneously.

Mandible fighters are found chiefly among the leaf-cutter ants, army ants and in the *Pheidole* genus, all three of which evolved soldiers or major-workers, these being particularly suited to mandible combat because of their very powerful mandibles. Further mandible fighters can be found among certain of the socially parasitic species, such as *Harpagoxenus, Strongylognathus* and *Polyergus.* Among leaf-cutters and the harvester genus *Pheidole,* mandible combat consists of quickly cutting off the extremities of the opponent. During this process, *Atta* keeps itself at a distance, suddenly bites and then quickly withdraws. Also among *Harpagoxenus* mandible combat is known where the opponent loses its extremities, similarly among *Camponotus* workers, which additionally use their poison, however. In the case of *Strongylognathus* and *Polyergus,* the mandibles have formed dagger-like weapons, with which they pierce the heads of their opponents.

New chemical weapons were developed in particular by the

Dolichoderines and the Formicines: the Dolichoderines *Tapinoma* and *Technomyrmex*, for example, spring up to their opponents in combat, quickly smear them with the secretion from their anal glands and then withdraw again (Maschwitz, 1975a). The anal gland was a new poison gland evolved by the Dolichoderines; the poison contained in the gland consists in *Tapinoma* of strong-smelling aliphatic ketones, and in other Dolichoderines of various lactones. The poisonous effect of the lactones is equal to that of DDT and is specifically effective with arthropods: lactones and ketones are hardly poisonous at all in the case of mammals, however (Cavill and Robertson, 1965; Weatherston, 1967). *Messor* workers may possibly represent a preliminary stage of the development of this type of chemical defence: their sting is also reduced and they smear their opponents in combat with foul-smelling faeces (Maschwitz, 1975a). A particularly interesting defence mechanism is that used by the Dolichoderine *Tapinoma nigerrimum*: in this species, the defence secretion consists of the ketones methylheptanone and propyl-isobutyl-keton, as well as dialdehyde iridodial, and is sprayed on to the opponent. The combination of chemicals polymerizes on the body of the opponent, thus rendering it immobile and also preventing the evaporation of the poisonous ketones (summary by Habermehl, 1976).

The chemical weapons of the Formicines consist mainly of the production of formic acid, which takes place, producing a 55% concentration, in the poison glands. A number of Formicines, the wood ants in particular, spray their opponents with poison from a distance. This gives them the advantage of being able to use their own weapons before being in range of their opponents. The formic acid penetrates the cuticula and the tracheal system into the inside of the insect opponents, and is an effective poison against them (Otto, 1960; Osman and Kloft, 1961). The Formicines also release the secretion of the Dufour's gland with the formic acid (Maschwitz, 1964a; Regnier and Wilson, 1968); this consists mostly of paraffins (Cavill and Williams, 1967), which serve in part as alarm substances, thus obtaining help from the area. These paraffins also strengthen the poisonous effect of the formic acid, by enabling it to spread over the opponent's cuticula more effectively (Regnier and Wilson, 1968). The fact that these paraffins are in themselves poisonous for insects, emerges from research conducted by Gilby and Cox (1963), who treated cockroaches (*Poriplaneta americana*) with the substances. It was first shown that the cockroaches were unable to move in a coordinated fashion, and also that in some cases they died. Among some ant species, such as *Crematogaster scutellaris,* the Dufour's gland has developed to such an extent at the expense of the poison gland, that it is now the only gland secreting the defensive substance (Maschwitz, 1975a).

Fig. 14.2   Ritual fight between workers of the species *Myrmecocystus mimicus*. (a) stilt walk of two workers which initially stand face to face; (b) the workers then stand side by side, running their antennae over the rear segment of their opponent (after Hölldobler, 1976b, adapted).

The new chemical weapons of the Dolichoderines and the Formicines do not always kill their opponents: frequently damage to the opponent is only localized, supporting the theory that these weapons were originally developed for competitive combat within the ants' own family. In this case it is not a question of capturing the opponent, but only of excluding it as a competitor.

Combat between colonies of *Myrmecocystus mimicus* proceed entirely without harm to the opponents, as was observed by Hölldobler (1976b) in south-west USA. Neighbouring colonies of this species occasionally cut across each other's boundaries when the areas in which they look for food happen to overlap. This leads to a territorial struggle, which in this species usually takes place by means of ritualized fighting. Hundreds of ants from both colonies in the disputed area come to these fights, which may continue for several days. The battles themselves usually only involve a few workers, which run at each other on their extended legs, their rear segments raised (Fig. 14.2a) and then stand side by side with their rear segments raised still higher. In this position, they try to bend the rear segment over in the direction of the opponent, at the same time running their antennae over the opponent's rear segment (Fig. 14.2b). This forms the only bodily contact between the participants. After 10-30 s, one of the two opponents usually surrenders. Where the two *Myrmecocystus* colonies are of different sizes, however, the ritual fights do not usually occur: after short and very numerous fights, which mostly involve the killing of opponents, the smaller of the two colonies is usually overrun and the nest plundered. The queen of the smaller colony is killed and all larvae, pupae and freshly hatched workers from the colony brought into the nest of the victors.

Other species avoid altercations which would be dangerous to the life of the ants, not by ritual fighting, but by quickly offering food to their attackers, thus pacifying them, as has been observed by Bhatkar and Kloft (1977) amongst numerous fire ant species (*Solenopsis*) and the harvester species *Pogonomyrmex dentata*. A similar exchange of food has also been observed taking place across species boundaries, among *Formica* and *Lasius* species, evidently occurring only between members of the same subfamily (Bhatkar, pers. inf.). Other species avoid dangerous encounters by playing dead, remaining motionless for several minutes, as has been observed by Schumacher and Whitford (1974) amongst *Trachymyrmex smithi neomexicanus*.

Territorial altercations also frequently proceed in a bloodless fashion between colony-founding females, and newly inseminated females and the members of already established colonies of the harvester species *Pogonomyrmex*. If a young *Pogonomyrmex badius* female in search of her own nesting place happens to cross into the territory of a colony of the same species, she is frequently not killed, but only carried to the territorial boundary and then released again. This method of removing increasing competition from their own territory in a bloodless manner was also found by Hölldobler (1976a) to exist among *Pogonomyrmex rugosa* and *Pogonomyrmex maricopa*. Even a *Pogonomyrmex barbatus* queen that forces into the founding chamber of another barbatus queen will not be attacked so as to endanger her life: the occupier of the founding chamber rather comes out and takes the alien queen about 3 m away from the chamber. She then returns to her own chamber (Hölldobler, 1976a).

Apart from the poison gland, Dufour's gland and the anal gland there are also, among some ant species, still further defence glands. The metathoracal glands, which produce antibiotics amongst almost all ant species, may thus also be brought into play against macro-organisms. The metathoracal gland of *Crematogaster inflata,* for example, is noticeably large, and its secretion visible from the outside. When the reservoir is full, the white secretion produced by the glands can be seen through the cuticle, giving the thorax of the animals a striking yellow colour (Fig. 14.3a). When they are fighting – mostly with other ants – the *Crematogaster inflata* workers release the extremely sticky secretion

Fig. 14.3   Crematogaster inflata worker (a), the whitish secretion of its metathoracal gland showing through its cuticula (dart glands); (b) *C. inflata* worker releasing the metathoracal gland secretion (S) (Photos — U. Maschwitz).

of these glands from both sides of the thorax (Fig. 14.3b). As soon as an opponent comes into contact with this secretion, it sticks to the animal, on occasion sticking its legs together. In addition, the secretion also raises the alarm to the members of the *Crematogaster inflata* colony, thus obtaining reinforcements. Any secretion in the glands which has not been used can be withdrawn again by the *Crematogaster inflata* workers. This ability, to actively release and withdraw the secretion of the metathoracal glands, is not so far known to exist in any other ant species (Maschwitz, 1974).

A further development of fighting technique is displayed by the Formicine subfamily: as a result of more effective weapons, their poison glands and thus also the production of formic acid, have reduced. This is the case for *Polyergus rufescens,* which produces only 1/40 of the quantity of formic acid produced by other Formicines (Stumper, 1960), and also for *Lasius fuliginosus,* which in comparison with *Lasius niger,* has also reduced its production (Stumper, after Maschwitz, unpublished). Whereas for *Polyergus* a very effective mandible fighting has taken the place of formic acid as a chemical weapon, in the case of *Lasius fuliginosus* a poison is produced in the mandible glands. These glands are so enlarged, that they constitute almost half the contents of the head. In addition to a number of other substances, the strong-smelling β(4,8-dimethylnon-3,7-dimethyl)-furan (dendrolasin) is produced (Pavan, 1956; Quilico *et al.*, 1956; Bernardi *et al.*, 1967). Dendrolasin is harmless to humans, but proves to be a strong poison for ants. This leads us to assume that this chemical weapon was not developed for use against predatory mammals, but specifically for combat with species related to themselves (Maschwitz, 1975a).

Also in *Camponotus saundersi*, which was investigated by Maschwitz (1974) in Malaysia, the mandible glands have changed their function to become used in fighting. The mandible glands of this species are so greatly enlarged that they extend the full length of the whole animal, as far as the tip of the rear segment (Fig. 14.4). In combat with other ants,

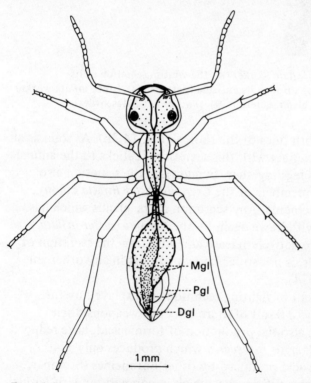

Fig. 14.4   Camponotus species close to *C. saundersi;* schematic representation of the paired mandible glands extending to the end of the gaster (Mgl); Pgl. formic acid producing poison gland; Dgl. Dufour's gland (from Maschwitz and Maschwitz, 1974).

the *Camponotus saundersi* workers use their weapons in a self-destructive way: by means of muscle contraction, they burst their rear segment, thus releasing the viscous, sticky secretion of the mandible gland. This secretion sticks the fighting animals to each other, thus rendering the opponent incapable of combat. This autothysia does not afford any advantage to the exploded ant itself, but it does to the total colony and population (Maschwitz and Maschwitz, 1974).

Summarizing the evolution of ant weapons, a complicated weapon, the stinging apparatus, developed from what was originally ovipositing apparatus. Whilst the cuticular parts of the stinging apparatus reformed, rather more effective chemical weapons developed out of the glands which had previously produced poisonous proteins. The chemical weapons were even effective across distances and could simultaneously alarm nest mates. In the Formicines, only formic acid is produced in the poison glands. The Dufour's gland also has a defence function, in that it reinforces the poisonous effect of the formic acid and also raises the alarm. In *Crematogaster scutellaris* the Dufour's gland is the sole poison gland. *Lasius fuliginosus* and *Polyergus rufescens* have reduced formic acid producing poison glands. In place of these, *Polyergus* has perfected mandible combat, whilst in the case of *Lasius (Dendrolasius) fuliginosus*, the mandible gland has become a defence gland and produces Dendrolasin.

For human beings, the stinging ants are usually much more unpleasant than those species where the sting is reduced. The new weapons which have been mentioned are, at least among the Central European species, not a source of fear to Man in any way, since mandible combat is useless against his tough skin and the chemical weapons are not poisonous for him. It should be quite clear in view of this, that these weapons were not developed to be used against Man and other large mammals, and that there are no universal weapons for ants, but only specific weapons to be used against specific enemies. The long-term advantage belongs to those species which are successful in maintaining their own in the face of their chief enemies, which are, in the majority of cases, probably their own species or close relatives.

Bibliography

ABE, T., 1971: *Jap. J. Ecol.*, **20**, 219–230.

ADLERZ, G., 1884: *Ófversign af kongl. Vetensk. Akad. Förhandl.*, **8**, 43–64.

ADLERZ, G., 1886: *Bihang Svensk. Vetensk. Akad. Handl.*, **11**, 1–320.

ADLERZ, G., 1896: *Bihang Svensk. Vetensk. Akad. Handl.*, **21**, 1–76.

AKRE, R. D., R. L. TORGERSON, 1969: *Pan Pacific Entomol.*, **45**, 269–281.

ALEXANDER, R. D., 1974: *Am. Rev. Ecol. Syst.*, **5**, 325–383.

ALPATOR, W. W., Z. G. PALENITSCHKO, 1925: *Rev. Zool. Russe.*, **5**, 109–116.

ALPERT, G. D., R. D. AKRE, 1973: *Ann. Entomol. Soc. Am.*, **66**, 753–760.

ANDRASFALVY, A., 1961: *Ins. soc.*, **8**, 299–310.

ARNOLD, G., 1914: *Proc. Rhod. Sci. Assoc.*, **13**, 25–32.

ARNOLD, G., 1915: *Ann. South Afric. Mus.*, **14**, 1–766.

ARNOLDI, K. V., 1928: *Zool. Anz.*, **89**, 299–310.

ARNOLDI, K. V., 1930: *Zool. Anz.*, **91**, 267–283.

ARNOLDI, K. V., 1932: *Z. Morph. Okol. Tiere*, **24**, 319–326.

AUCLAIR, J. L., 1963: *Ann. Rev. Entomol*, **8**, 439–490.

AUTRUM, H., 1936: *Z. vergl. Physiol.*, **23**, 332–373.

AUTRUM, H., W. SCHNEIDER, 1948: *Z. vergl. Physiol.*, **31**, 77–88.

AYRE, G. L., M. S. BLUM, 1971: *Physiol. Zool.*, **44**, 77–83.

BARBIER, M., E. LEDERER, 1960: *C. R. Acad. Sc. Paris*, **25**, 4467–4469.

BARLIN, M. R., M. S. BLUM, J. M. BRAND, 1976: *J. Ins. Physiol.*, **22**, 839–844.

BARONI-URBANI, C., 1969: *Boll. Soc. Entomol. Ital.*, **99–101**, 132–168.

BAZIRE-BENAZET, M., 1957: *C. R. Acad. Sc. Paris*, **244**, 1277–1280.

BELT, T., 1928: *The naturalist in Nicaragua*. London: Dent.

BENOIS, A., 1972 a: *Ann. Zool. Ecol. anim.*, **4**, 325–351.

BENOIS, A., 1972 b: *Ins. soc.*, **19**, 111–129.

BENZING, D. H., 1970: *Bull. Torrey Bot. Club*, **97**, 109–115.

BEQUAERT, J., 1922: *Bull. Mus. Nat. Hist.*, **45**, 333–583.

BERGSTRÖM, G., J. LÖFQVIST, 1968: *J. Ins. Physiol.*, **14**, 995–1011.

BERGSTRÖM, G., J. LÖFQVIST, 1970: *J. Ins. Physiol.*, **16**, 2353–2375.

BERGSTRÖM, G., J. LÖFQVIST, 1973: *J. Ins. Physiol.*, **19**, 877–907.

BERNARD, F., 1968: *Les fourmis (Hymenoptera, Formicidae) d'Europe occidentale et septentrionale. Faune de l'Europe et du Bassin Méditerrané*, Paris: Masson et Cie.

BERNARDI, C. D., D. CARDANI, D. GHIRINGHELLI, A. SELVA, A. BAGGINI, M. PAVAN, 1967: *Tetrahedron Letters*, **40**, 3893–3896.

BERNSTEIN, R. A., 1974: *The Am. Natural.*, **108**, 490–498.

BETHE, A., 1932: *Naturw.*, **20**, 177–181.

BETHE, J. D., 1898: *Pflügers Arch. ges. Physiol.*, **70**, 15–100.

BHATKAR, A., W. H. WHITCOMB, 1970: *The Florida Entomol.*, **53**, 229–232.

BHATKAR, A., W. H. WHITCOMB, W. F. BUREN, P. CALLAHAN, T. CARLYSLE, 1972: *Environm. Entomol.*, **1**, 274–279.

BHATKAR, A. P., W. J. KLOFT, 1977: *Nature*, **265**, 140–142.

BIER, K., 1954 a: *Ins. soc.*, **1**, 7–19.

BIER, K., 1954 b: *Biol. Zbl.*, **73**, 170–190.

BIER, K., 1956: *Ins. soc.*, **3**, 177–184.

BIER, K., 1958: *Erg. Biol.*, **20**, 97–126.

BINGHAM, C. T., 1903: *The fauna of British India, including Ceylon and Burma. Hymenoptera, Vol. II.*, London: Taylor & Francis.

BISCHOFF, H., 1927: *Biologie der Hymenopteren*. Berlin: Springer.

BLUM, M. S., 1966: *Proc. R. Entomol. Soc. London*, **41**, 155–160.

BLUM, M. S., 1969: *Ann. Rev. Entomol.*, **14**, 57–80.

BLUM, M. S., 1970: 61–94. In: BEROZA, ed., *Chemicals controlling insect behavior*. New York: Academic Press.

BLUM, M. S., J. B. BYRD, J. R. TRAVIS, J. F. WATKINS, F. G. GEHLBACH, 1971: *Comp. Biochem. Physiol.*, **38 B**, 103–107.

BLUM, M. S., E. M. CREWE, J. M. PASTEELS, 1971: *Ann. Entomol. Soc. Am.*, **64**, 975–976.

BLUM, M. S., S. L. WARTER, 1966: *Ann. Entomol. Soc. Am.*, **59**, 774–779.

BLUM, M. S., E. O. WILSON, 1964: *Psyche*, **71**, 28–31.

BOCH, R., D. A. SHEARER, 1965: *Nature*, **206**, 530.

BOCH, R., D. A. SHEARER, 1967: *Z. vergl. Physiol.,* **54**, 1–11.
BOCH, R., D. A. SHEARER, B. C. STONE, 1962: *Nature,* **195**, 1018–1020.
BORGMEIER, T., 1937: *Rev. Entomol.,* **7**, 129–134.
BORGMEIER, T., 1955: *Stud. Entomol.,* **3**, 1 – 716.
BOSSERT, W. H., E. O. WILSON, 1963: *J. theor. Biol.,* **5**, 443–469.
BOYD, N. D., M. M. MARTINI, 1975 a: *J. Ins. Physiol.,* **21**, 1815–1820.
BOYD, N. D., M. M. MARTINI, 1975 b: *Ins. Biochem.,* **5**, 619–635.
BRADSHAW, J. W. S., R. BAKER, P. E. HOWSE, 1975: *Nature,* **258**, 230–231.
BRANDER, J., 1959: *Die chemische Untersuchung einiger Gespinste von Spinnen und Ameisen.* Diss. Nat. Fak. Würzburg.
BRIAN, M. V., 1954: *Ins. soc.,* **1**, 101–122.
BRIAN, M. V., 1955 a: *J. Anim. Ecol.,* **24**, 336–351.
BRIAN, M. V., 1955 b: *Ins. soc.,* **2**, 1–25.
BRIAN, M. V., 1956 a: *Ins. soc.,* **3**, 369–394.
BRIAN, M. V., 1956 b: *J. Anim. Ecol.,* **25**, 319–337.
BRIAN, M. V., 1956 c: *J. Anim. Ecol.,* **25**, 339–347.
BRIAN, M. V., 1957: *Ann. Rev. Entomol.,* **2**, 107–120.
BRIAN, M. V., 1963: *Ins. soc.,* **10**, 91–102.
BRIAN, M. V., 1965 a: *Social insect populations.* New York: Academic Press.
BRIAN, M. V., 1965 b: *Symp. Zool. Soc. London,* **14**, 13–38.
BRIAN, M. V., 1973 a: *Ins. soc.,* **20**, 87–102.
BRIAN, M. V., 1973 b: *J. Anim. Ecol.,* **42**, 37–53.
BRIAN, M. V., 1973 c: *Physiol. Zool.,* **46**, 245–252.
BRIAN, M. V., 1974 a: *J. Ins. Physiol.,* **20**, 1351–1365.
BRIAN, M. V., 1974 b: L. In: *Sozialpolymorphismus bei Insekten. Probleme der Kastenbildung im Tierreich,* hrsg. G. H. Schmidt. Stuttgart: Wissenschaftliche Verlagsgesellschaft.
BRIAN, M. V., 1975: *Anim. Behav.,* **23**, 745–756.
BRIAN, M. V., M. S. BLUM, 1969: *J. Ins. Physiol.,* **15**, 2213–2223.
BRIAN, M. V., A. D. BRIAN, 1955: *Evolut.,* **9**, 280–290.
BRIAN, M. V., C. A. H. CARR, 1960: *J. Ins. Physiol.,* **5**, 81–94.
BRIAN, M. V., B. M. DOWNING, 1958: *Proc. Xth Int. Congr. Entomol. Montreal,* **2**, 539–540.
BRIAN, M. V., J. HIBBLE, 1963: *J. Ins. Physiol.,* **9**, 25–34.
BRIAN, M. V., J. HIBBLE, 1964: *Ins. soc.,* **11**, 223–238.
BRIAN, M. V., A. F. KELLY, 1967: *Ins. soc.,* **14**, 13–24.

BROWER, H., 1966: *Am. Midl. Natural.,* **75**, 530–534.

BROWN, W. L., 1954: *Ins. soc.,* **1**, 21–31.

BROWN, W. L., 1955: *Ins. soc.,* **2**, 181–186.

BROWN, W. L., 1960: *Psyche,* **66**, 25–27.

BROWN, W. L., 1967: *Psyche,* **74**, 331–339.

BROWN, W. L., 1968: *Am. Nat.,* **102**, 188–191.

BROWN, W. L., T. EISNER, R. H. WHITTAKER, 1970: *Bio Science,* **20**, 21–22.

BROWN, W. L., W. L. NUTTING, 1950: *Trans. Am. Soc.,* **75**, 113–132.

BROWN, W. L., E. O. WILSON, 1959: *Quart. Rev. Biol.,* **34**, 278–294.

BRUN, R., 1910: *Biol. Cbl.,* **30**, 529–545.

BRUN, R., 1913: *Biol. Cbl.,* **33**, 17–29.

BRUN, R., 1914: *Die Raumorientierung der Ameisen und das Orientierungsproblem im allgemeinen.* Jena: G. Fischer.

BRUN, R., 1952: *Mitt. Schweiz. Entomol. Ges.,* **25**, 73–86.

BÜCKMANN, D., 1962: *Naturw.,* **49**, 28–33.

BÜNZLI, G. H., 1935: *Mitt. Schweiz. Entomol. Ges.,* **16**, 453–593.

BURGETT, D. M., R. G. YOUNG, 1974: *Ann. Entomol. Soc. Am.,* **67**, 743–744.

BUSCHINGER, A., 1965: *Ins. soc.,* **12**, 327–334.

BUSCHINGER, A., 1966 a: *Ins. soc.,* **13**, 5–16.

BUSCHINGER, A., 1966 b: *Ins. soc.,* **13**, 311–322.

BUSCHINGER, A., 1966 c: *Ins. soc.,* **13**, 165–172.

BUSCHINGER, A., 1967: *Verbreitung und Auswirkungen von Mono- und Polygynie bei Arten der Gattung Leptothorax Mayr (Hymenoptera, Formicidae).* Diss. Univ. Würzburg.

BUSCHINGER, A., 1968 a: *Ins. soc.,* **15**, 89–104.

BUSCHINGER, A., 1968 b: *Experient.,* **24**, 297.

BUSCHINGER, A., 1968 c: *Ins. soc.,* **15**, 217–225.

BUSCHINGER, A., 1968 d: *Bayer. Tierw.,* **1**, 115–128.

BUSCHINGER, A., 1970: *Biol. Zbl.,* **89**, 273–299.

BUSCHINGER, A., 1971 a: *Zool. Anz.,* **186**, 242–248.

BUSCHINGER, A., 1971 b: *Zool. Anz.,* **187**, 184–191.

BUSCHINGER, A., 1971 c: *Zool. Anz.,* **186**, 47–59.

BUSCHINGER, A., 1971 d: *Bonn. Zool. Beitr,* **22**, 321–331.

BUSCHINGER, A., 1972 a: *Naturw.,* **59**, 313–314.

BUSCHINGER, A., 1972 b: *Zool. Anz.,* **189**, 169–179.

BUSCHINGER, A., 1973 a: *Zool. Anz.,* **190**, 63–66.

BUSCHINGER, A., 1973 b: *Proc. VII. Congr. IUSSI, London,* 50–55.

BUSCHINGER, A., 1974: *Ins. soc.,* **21**, 133–144.

BUSCHINGER, A., 1975: *Naturw.*, **62**, 239.

BUSCHINGER, A., 1976 a: *Ins. soc.*, **23**, 205–214.

BUSCHINGER, A., 1976 b: *Ins. soc.*, **23**, 215–226.

BUSCHINGER, A., W. KLOFT, 1969: *Anz. Schädlingsk. Pflanzensch.*, **42**, 49–53.

BUSCHINGER, A., W. KLOFT, 1973: *Forschungsb. L. Nordrh.-Westf.*, Nr. 2306, Opladen: Westdeutscher Verlag.

BUSCHINGER, A., M. PETERSEN, 1971: *Anz. Schädlingsk., Pflanzensch.*, **44**, 103–106.

BUSCHINGER, A., U. WINTER, 1975: *Ins. soc.*, **22**, 333–362.

BUSCHINGER, A., U. WINTER, 1978: *Ins. soc.*, **25**, 63–78.

BUTLER, C. G., R. K. CALLOW, J. R. CHAPMAN, 1964: *Nature*, **201**, 733.

BUTLER, C. G., J. SIMPSON, 1958: *Proc. Roy. Entomol. Soc. London (A)*, **33**, 120–122.

BUTTEL-REEPEN, H. v., 1900: *Biol. Cbl.*, **20**, 1–82.

BÜTTNER, K., 1974 a: *Waldhyg.*, **10**, 129–140.

BÜTTNER, K., 1974 b: *Waldhyg.*, **10**, 161–182.

CADWELL, L. L., 1973: *Am. Midl. Nat.*, **89**, 446–448.

CAGNIANT, H., 1970: *Ins. soc.*, **17**, 39–48.

CAGNIANT, H., 1973: *C. R. Acad. Sc. Paris, Ser. D.*, **277**, 2197–2198.

CALLOW, P. K., J. R. CHAPMAN, P. N. PATON, 1964: *J. Apicult. Res.*, **3**, 77–89.

CALLOW, P. K., N. C. JOHNSTON, 1960: *Bee World*, **41**, 152–153.

CAMMAERTS-TRICOT, M.-C., 1973: *J. Ins. Physiol.*, **19**, 1299–1315.

CAMMAERTS-TRICOT, M.-C., 1974 a: *J. comp. Physiol.*, **88**, 373–382.

CAMMAERTS-TRICOT, M.-C., 1974 b: *Behav.*, **50**, 111–112.

CAMMAERTS-TRICOT, M.-C., 1974 c: *Ins. soc.*, **21**, 235–248.

CAMMAERTS-TRICOT, M.-C., J.-C. VERHAEGHE, 1974: *Ins. soc.*, **21**, 275–282.

CARR, C. A. H., 1962: *Ins. soc.*, **9**, 197–211.

CARTHY, J. D., 1951 a: *Behav.*, **3**, 275–303.

CARTHY, J. D., 1951 b: *Behav.*, **3**, 304–318.

CASEVITZ-WEULERSSE, 1972: *Bull. Soc. Entomol. France*, **77**, 12–19.

CASNATI, R. A., E. RICCA, M. PAVAN, 1967: *Chim. Ind. (Milano)*, **49**, 57–61.

CAVILL, G. W. K., P. J. WILLIAMS, 1967: *J. Ins. Physiol.*, **13**, 1097–1103.

CAVILL, W. K., P. L. ROBERTSON, 1965: *Science*, **149**, 1337–1345.

CHAUVIN, R., 1952: *Behav.*, **4**, 190–201.

CHERRETT, J. M., 1972: *J. Anim. Ecol.*, **41**, 647–660.

CLARK, J., 1934: *Mém. Nat. Mus. Melbourne*, **8**, 5–20.

CLARK, J., 1951: *The Formicidae of Australia. Volume I. Subfamily Myrmeciinae.* Melbourne: Commonwealth Scientific and Industr. Res. Org.

CLARK, W. H., P. L. COMANOR, 1973: *Am. Midl. Natural.*, **90**, 467–474.

CLOUDSLEY-THOMPSON, J. L., J. H. P. SANKEY, 1958: *Entomol. Month. Mag.*, **94**, 43–47.

COHIC, F., 1948: *Rev. Franc. Entomol.*, **14**, 229–276.

COLLART, A., 1925: *Bull. Zool. Cong.*, **2**, 26–28.

COLLINGWOOD, C. A., 1956: *Entomol. Month. Mag.*, **92**, 197.

COLLINGWOOD, C. A. 1958: *Proc. Roy. Entomol. Soc. London (A)*, **33**, 65–75.

COLOMBEL, P., 1970 a: *Ins. soc.*, **17**, 183–198.

COLOMBEL, P., 1970 b: *Ins. soc.*, **17**, 199–204.

COLOMBEL, P., 1971 a: *C. R. Acad. Sc. Paris*, **272**, 970–972.

COLOMBEL, P., 1971 b: *C. R. Acad. Sc. Paris*, **272**, 2710–2712.

COLOMBEL, P., 1971 c: *Ann. fac. Sc. Yaoundé*, **6**, 53–71.

COLOMBEL, P., 1971 d: *Ins. soc.*, **18**, 183–198.

COLOMBEL, P., 1972 a: *Ins. soc.*, **19**, 171–194.

COLOMBEL, P., 1972 b: *Biol. Gabon. (Périgueux)*, **8**, 345–367.

COLOMBEL, P., 1972 c: *Biol. Gabon. (Périgueux)*, **8**, 369–381.

CORNETZ, V., 1913: *Les explorations et les voyages des fourmis.* Paris: Flammarion.

CRAWLEY, W. C., 1909: *Entomol. Month. Mag.*, **20**, 94–99.

CRAWLEY, W. C., 1911: *Trans. Entomol. Soc. London*, 657–663.

CRAWLEY, W. C., H. DONISTHORPE, 1912: *Trans. II. Entomol. Congr.*, 73–86.

CREIGHTON, W. S., 1929: *Psyche*, **36**, 48–50.

CREIGHTON, W. S., 1950: *Bull. Mus. Comp. Zool.*, **104**, 1–585.

CREIGHTON, W. S., 1963: *Psyche*, **70**, 133–143.

CREIGHTON, W. S., M. P. CREIGHTON, 1959: *Psyche*, **66**, 1–2.

CREIGHTON, W. S., R. E. GREGG, 1954: *Psyche*, **61**, 41–57.

CREWE, R. M., M. S. BLUM, 1970: *Z. vergl. Physiol.*, **70**, 363–373.

CROZIER, R. H., 1970: *Can. J. Gen. Cytol.*, **12**, 109–128.

DE BACH, P., 1951: *J. Econ. Entomol.*, **44**, 443–447.
DEJEAN, A., L. PASSERA, 1974: *Ins. soc.*, **21**, 343–356.
DELAGE, B., 1962: *Ins. Soc.*, **9**, 137–143.
DELAGE-DARCHEN, B., 1972 a: *Ins. soc.*, **19**, 213–226.
DELAGE-DARCHEN, B., 1972 b: *Ins. soc.*, **19**, 259–278.
DELAGE-DARCHEN, B., 1974: *Ins. soc.*, **21**, 13–34.
DÉLYE, G., 1957: *Ins. soc.*, **4**, 77–82.
DÉLYE, G., 1971: *Ins. soc.*, **18**, 15–20.
DÉLYE, G., 1974: *Ins. soc.*, **21**, 369–380.
DOBRZANSKA, J., 1959: *Acta Biol. Exp.*, **19**, 57–81.
DOBRZANSKA, J., 1966: *Acta Biol. Exp.*, **26**, 193–213.
DOBRZANSKA, J., 1973: *Acta Neurobiol. Exp.*, **33**, 597–622.
DOBRZANSKA, J., J. DOBRZANSKI, 1960: *Ins. soc.*, **7**, 1–8.
DOBRZANSKI, J., 1961: *Acta Biol. Exp.*, **21**, 53–73.
DOBRZANSKI, J., 1965: *Acta Biol. Exp.*, **25**, 59–71.
DOBRZANSKI, J., 1966: *Acta Biol. Exp.*, **26**, 71–78.
DOFLEIN, F., 1905: *Biol. Cbl.*, **25**, 497–507.
DONISTHORPE, H., 1927: *British ants.* London: Routledge.
DOUGLAS, A., W. L. BROWN, 1959: *Ins. soc.*, **6**, 13–19.
DOWNHOWER, J. F. 1975: *Biotropica*, **7**, 59–62.
DUELLI, P., 1974: *Polarisationsmusterorientierung der Wüstenameise Cataglyphis bicolor (Formididae, Hymenoptera).* Diss. Univ. Zürich.
DUELLI, P., 1975: *J. comp. Physiol.*, **102**, 43–56.
DUELLI, P., R. WEHNER, 1973: *J. comp. Physiol.*, **86**, 37–53.
DUMPERT, K., 1972 a: *Z. vergl. Physiol.*, **76**, 403–425.
DUMPERT, K., 1972 b: *Z. Morph. Tiere*, **73**, 95–116.
DUVIARD, D., P. SEGEREN, 1974: *Ins. soc.*, **21**, 191–212.

EHRHARDT, H. J., 1962: *Naturw.*, **49**, 524–525.
EHRHARDT, H. J., 1970: *Die Bedeutung von Königinnen mit steter arrhenotoker Parthenogenese für die Männchenerzeugung in den Staaten von Formica polyctena Foerster (Hym., Form.).* Diss. Univ. Würzburg.
EHRHARDT, S., 1931: *Z. Morph. Ökol. Tiere*, **20**, 755–812.
EIBL-EIBESFELDT, I., E. EIBL-EIBESFELDT, 1967: *Z. Tierpsych.*, **24**, 278–281.
EIDMANN, H., 1926: *Z. vergl. Physiol.*, **3**, 776–826.
EIDMANN, H., 1927: *Biol. Zbl.*, **47**, 537–556.

EIDMANN, H., 1928 a: *Z. vergl. Physiol.,* 7, 39—55.

EIDMANN, H., 1928 b: *Z. angew. Entomol.,* 14, 229—253.

EIDMANN, H., 1929 a: *Z. Forsch. Fortschr.,* 5, 41—42.

EIDMANN, H., 1929 b: *Zool. Anz.,* 82, 99—114.

EIDMANN, H., 1931: *Biol. Zbl.,* 51, 657—677.

EIDMANN, H., 1932: *Z. Morph. Ökol. Tiere,* 25, 154—183.

EIDMANN, H., 1943: *Z. Morph. Ökol. Tiere,* 39, 217—275.

EISNER, T., 1957: *Bull. Mus. Comp. Zool.,* 116, 439—490.

EISNER, T., W. L. BROWN, 1958: *Proc. 10. Int. Congr. Entomol.,* 2, 503—508.

EISNER, T., E. v. TASSEL, J. E. CARREL, 1967: *Science,* 158, 1471—1473.

EISNER, T., E. O. WILSON, 1952: *Psyche,* 59, 47—60.

EISNER, T., E. O. WILSON, 1958: *Proc. 10. Intern. Congr. Entomol.,* 2, 509—513.

ELMES, G. W., 1973 a: *J. Anim. Ecol.,* 42, 761—771.

ELMES, G. W., 1973 b: *The Entomol.,* 106, 133—136.

ELMES, G. W., 1974: *Oecologia,* 15, 337—343.

ELMES, G. W., 1976: *Ins. soc.,* 23, 3—22.

ELTON, C., 1932: *J. Anim. Ecol.,* 1, 69—76.

EL-ZIADY, S., J. S. KENNEDY, 1956: *Proc. Roy. Entomol. Soc. London (A),* 31, 61—65.

EMERY, C., 1896: *C. R. Séances Trois. Congr. Int. Zool., Leyden, 1895,* 395—410.

EMERY, C., 1910—1925: Formicidae: Dorylinae, Ponerinae, Dolichoderinae, Formicinae. In: P. Wytsman, *Genera Insectorum, fasc.* 102 (1910), fasc. 118 (1911), fasc. 137 (1912), fasc. 174 (1921—1922), fasc. 183 (1925). Brussels: V. Verteneuil et L. Desmet.

EMMERT, W., 1969: *Roux' Arch.,* 162, 97—113.

ENGELHARDT, W., 1970: *Zool. Anz.,* 185, 317—334.

ESCHERICH, K., 1917: *Die Ameise.* Braunschweig: F. Vieweg u. Sohn.

EVANS, H. E., M. J. W. EBERHARD, 1970: *The Wasps.* Ann Arbor: Univ. Michigan Press.

EVANS, W. C., D. LESTON, 1971: *Bull. Entomol. Res.,* 61, 357—362.

FABER, W., 1967: *Pflanzensch.-Ber.,* 36, 73—107.

FABER, W., 1969: *Pflanzensch.-Ber.,* 39, 39—100.

FALKE, J., 1968: *Substanzen aus der Mandibeldrüse der Männchen von Camponotus herculeanus.* Diss. Univ. Heidelberg.

FELLERS, J. H., G. M. FELLERS, 1976: *Science,* 192, 70—72.

FIELDE, A. M., 1901: *Proc. Acad. Nat. Sc. Philadelphia,* **53,** 521–544.
FIELDE, A. M., 1903: *Biol. Bull., Marine Biol. Lab., Woods Hole,* **5,** 320–325.
FLETCHER, D. J. C., J. M. BRAND, 1968: *J. Ins. Physiol.,* **14,** 783–788.
FOERSTER, E., 1912: *Zool. Jhb. Anat. Ontog.,* **34,** 347–380.
FOREL, A., 1869: *Mitt. Schweiz. Entomol. Ges.,* **3.**
FOREL, A., 1874: *Les fourmis de la Suisse.* Zürich: Societé Helvétique des Sciences Naturelles. Revidierte und korrigierte Fassung, Imprimerie Coopérative, La Chaux-de-fonds (1920).
FOREL. A., 1898: *Bull. Soc. Vaudoise Sci. Nat.,* **34,** 380–384.
FOREL, A., 1901: *Ann. Soc. Entomol. Belg.,* **45,** 389–398.
FOREL, A., 1902: *J. Psychol. Neurol.,* **1,** 99–110.
FOREL, A., 1905: *Biol. Cbl.,* **25,** 170–181.
FOREL, A., 1921–1923: *Le monde social des fourmis du globe comparée à celui de l'homme.* Geneva: Libraire Kundig, 5 Bde.
FRANCOEUR, A., 1968: *H. F. Nat. Cand.,* **95,** 727–730.
FREE, J. B., J. SIMPSON, 1968: *Z. vergl. Physiol.,* **61,** 361–365.
FREELAND, J., 1958: *Myrmecia. Austr. J. Zool.,* **6,** 1–18.
FRISCH, K. v., 1950: *Experientia,* **6,** 210–221.
FRISCH, K. v., 1965: *Tranzsprache und Orientierung der Bienen.* Berlin–Heidelberg–New York: Springer.
FRISCH, K. v., 1974: *Tiere als Baumeister.* Frankfurt/M.–Berlin– Wien: Ullstein.

GALLÉ, L., 1973: *Acta Biol. Szeged,* **19,** 139–142.
GASPAR, C., 1967: *Ins. soc.,* **14,** 183–190.
GEHLBACH, F. R., J. F. WATKINS, H. W. RENO, 1968: *Bioscience,* **18,** 784–785.
GHIGI, A., 1951: *La vita degli animali.* Turin: Chione tip. – ed. Tornise.
GILBY, A. R., M. E. COX, 1963: *J. Ins. Physiol.,* **9,** 671–681.
GLANCEY, B. M., C. E. CRAIG, P. M. BISHOP, B. B. MARTIN, 1970: *Nature, London,* **226,** 863–864.
GOETSCH, W., 1930: *Z.Morph. Ökol. Tiere,* **16,** 371–452.
GOETSCH, W., 1934: *Z. Morph. Ökol. Tiere,* **28,** 319–401.
GOETSCH, W., 1937: *Naturw.,* **25,** 803–808.
GOETSCH, W., 1940: *Vergleichende Biologie der Insektenstaaten.* Leipzig: Akademische Verlagsgesellsch. Becker und Erler.
GOETSCH, W., 1953: *Die Staaten der Ameisen.* Berlin–Heidelberg–

New York: Springer.

GOETSCH, W., B. KÄTHNER, 1938: *Z. Morph. Ökol. Tiere, 33*, 201–260.

GÖRNER, P., 1973: *Fortschr. Zool., 21*, 20–45.

GÖSSWALD, K., 1930: *Z. wiss. Zool., 136*, 464–484.

GÖSSWALD, K., 1933: *Z. wiss. Zool., 144*, 262–288.

GÖSSWALD, K., 1934: *Forsch. Fortschr., 279–280*.

GÖSSWALD, K., 1938: *Z. wiss. Zool., 151*, 101–148.

GÖSSWALD, K., 1951: *Die Rote Waldameise im Dienst der Waldhygiene*. Lüneburg: Metta Kinau.

GÖSSWALD, K., 1953: *Mitt. Schweiz. Entomol. Ges., 26*, 81–128.

GÖSSWALD, K., 1954: *Unsere Ameisen*. Stuttgart: Kosmos.

GÖSSWALD, K., K. BIER, 1953 a: *Naturw., 40*, 38–39.

GÖSSWALD, K., K. BIER, 1953 b: *Zool. Anz., 151*, 126–134.

GÖSSWALD, K., K. BIER, 1954 a: *Ins. soc., 1*, 229–246.

GÖSSWALD, K., K. BIER, 1954 b: *Ins. soc., 1*, 305–318.

GÖSSWALD, K., K. BIER, 1957: *Ins. soc., 4*, 335–348.

GÖSSWALD, K., W. KLOFT, 1956: *Waldhygiene, 1*, 200–202.

GÖSSWALD, K., W. KLOFT, 1958: *Proc. 10. Intern. Congr. Entomol., London, 2*, 543.

GÖSSWALD, K., W. KLOFT, 1960 a: *Zool. Beitr. N. F., 5*, 519–556.

GÖSSWALD, K., W. KLOFT, 1960 b: *Entomophaga, 5*, 33–41.

GÖSSWALD, K., W. KLOFT, 1963: *Symp. Biol. Ital., Pavia, 12*, 9–14.

GÖSSWALD, K., G. H. SCHMIDT, 1960: *Ins. soc., 7*, 297–321

GOTWALD, W. H., 1969: *Memoir, 408*, 1–150.

GOTWALD, W. H., 1970: *Ann. Entomol. Soc. Am., 63*, 950–952.

GOTWALD, W. H., 1972: *Psyche, 79*, 348–356.

GOTWALD, W. H., 1974 a: *Ann. Entomol. Soc. Am., 67*, 877–886.

GOTWALD, W. H., 1974 b: *Bull. l'I. F. A. N., 36*, 705–713.

GOTWALD, W. H., J. LEVIEUX, 1972: *Ann. Entomol. Soc. Am., 65*, 383–396.

GOULD, W., 1747: *An account of English ants*. London: A. Millar.

GOUNELLE, E., 1900: *Bull. Soc. Entomol. France, 168–169*.

GRAY, B., 1971 a: *Ins. soc., 18*, 71–80.

GRAY, B., 1971 b: *Ins. soc., 18*, 81–94.

GRAY, B., 1974: *Ins. soc., 21*, 107–120.

GREEN, E. E., 1896: *Trans. Entomol Soc. London, 9*–10.

GREGG, E. E., 1972: *Can. Entomol., 104*, 1073–1091.

GREGG, R. E., 1942: *Ecology, 23*, 295–308.

HABERMEHL, G., 1976: *Gift- Tiere und ihre Waffen.*
Berlin—Heidelberg—New York: Springer.

HAMILTON, W. D., 1964: *J. Biol.,* 7, 1—52.

HANGARTNER, W., 1967: *Z. vergl. Physiol.,* 57, 103—136.

HANGARTNER, W., 1969 a: *Ins. soc.,* 16, 55—60.

HANGARTNER, W., 1969 b: *J. Ins. Physiol.,* 15, 1—4.

HANGARTNER, W., 1969 c: *Z. vergl. Physiol.,* 62, 111—120.

HANGARTNER, W., 1970: *Experient,* 26, 664—665.

HANGARTNER, W., S. BERNSTEIN, 1964: *Experient.,* 20, 392—393.

HANGARTNER, W., J. M. REICHSON, E. O. WILSON, 1970: *Anim. Behav.,* 18, 331—334.

HASKINS, C. P., 1970: In: L. R. ARONSON, E. TOBACH, D. S. LEHRMAN and J. S. ROSENBLATT eds., *Development and revolution of behavior.* 355—388. San Francisco: Freeman and Comp.

HASKINS, C. P., E. V. ENZMANN, 1945: *J. New York Entomol. Soc.,* 53, 263—277.

HASKINS, C. P., E. F. HASKINS, 1950: *Psyche,* 57, 1—9.

HASKINS, C. P., E. F. HASKINS, 1951: *Am. Midl. Natural.,* 45, 432—445.

HASKINS, C. P., E. F. HASKINS, 1955: *Ins. soc.,* 2, 115—126.

HASKINS, C. P., E. F. HASKINS, 1964: *Ins. soc.,* 11, 267—282.

HASKINS, C. P., E. F. HASKINS, 1974: *Psyche,* 81, 258—267.

HASKINS, C. P., R. M. WHELDEN, 1954: *Ins. soc.,* 1, 33—37.

HASKINS, C. P., R. M. WHELDEN, 1965: *Psyche,* 72, 87—112.

HAUSCHTECK, E., 1965: *Experient.,* 21, 323—325.

HEMMINGSEN, A. M., 1973: *Vidensk. Meddr. dansk. naturh. Foren,* 136, 49—56.

HENNIG, W., 1969: *Stammesgeschichte der Insekten.* Frankfurt/M.: W. Kramer.

HERMANN, H. R., 1968: *Ann. Entomol. Soc. Am.,* 61, 1315—1317.

HERMANN, H. R., M. S. BLUM, 1967: *Ann. Entomol. Soc. Am.,* 60, 661—668.

HERTER, K., 1923: *Biol. Zbl.,* 43, 282—285.

HERTER, K., 1924: *Z. vergl. Physiol.,* 1, 221—288.

HERTER, K., 1925: *Z. vergl. Physiol.,* 2, 226—232.

HERZIG, J., 1938: *Z. angew. Entomol.,* 24, 367—435.

HEYDE, K., 1924: *Biol. Zbl.,* 44, 623—654.

HINGSTON, R. W. G., 1927: *Proc. Roy. Entomol. Soc. London,* 2, 90—94.

HINGSTON, R. W. G., 1929: *Instinct and Intelligence.* New York: Mac Millan and Co.

HOHORST. B., 1972: *Ins. soc.,* **19**, 389–402.

HÖLLDOBLER, B., 1962: *Z. angew. Entomol.,* **49**, 337–352.

HÖLLDOBLER, B., 1966: *Z. vergl. Physiol.,* **52**, 430–455.

HÖLLDOBLER, B., 1967 a: *Z. vergl. Physiol.,* **56**, 1–21.

HÖLLDOBLER, B., 1967 b: *Zool. Anz.,* **31**, Suppl.-Band, 428–434.

HÖLLDOBLER, B., 1968: *Naturw.,* **55**, 397.

HÖLLDOBLER, B., 1969 a: *Science,* **166**, 757–758.

HÖLLDOBLER, B., 1969 b: *Verh. Dtsch. Zool. Ges. Würzburg,* 580–585.

HÖLLDOBLER, B., 1970 a: *Z. vergl. Physiol.,* **66**, 215–220.

HÖLLDOBLER, B., 1970 b: *Umschau Wiss. Techn.,* **70**, 663–669.

HÖLLDOBLER, B., 1971 a: *J. Ins. Physiol.,* **17**, 1497–1499.

HÖLLDOBLER, B., 1971 b: *Z. vergl. Physiol.,* **75**, 123–142.

HÖLLDOBLER, B., 1971 c: *Scientific Am.,* **224**, 86–93.

HÖLLDOBLER, B., 1973 a: *Encycl. Cin., Göttingen.*

HÖLLDOBLER, B., 1973 b: *Encycl. Cin., Göttingen.*

HÖLLDOBLER, B., 1973 c: *Oecologia,* **11**, 371–380.

HÖLLDOBLER, B., 1973 d: *Nova Acta Leopoldina,* **37**, 259–292.

HÖLLDOBLER, B., 1974: *Proc. Nat. Acad. Sci.,* **71**, 3274–3277.

HÖLLDOBLER, B., 1976 a: *Behav. Ecol. Sociobiol.,* **1**, 3–44.

HÖLLDOBLER, B., 1976 b: *Science,* **192**, 912–914.

HÖLLDOBLER, B., U. MASCHWITZ, 1965: *Z. vergl. Physiol.,* **50**, 551–568.

HÖLLDOBLER, B., M. MÖGLICH, U. MASCHWITZ, 1973: *Encycl. Cin., Göttingen.*

HÖLLDOBLER, B., E. O. WILSON, 1970: *Psyche,* **77**, 385–399.

HÖLLDOBLER, B., M. WÜST, 1973: *Z. Tierpsych.,* **32**, 1–9.

HÖLLDOBLER, K. 1923: *Aus der Heimat,* **36**, 114–116.

HÖLLDOBLER, K., 1928: *Biol. Zbl.,* **48**, 129–142.

HÖLLDOBLER, K., 1936: *Biol. Zbl.,* **56**, 230–248.

HÖLLDOBLER, K., 1938 a: *Z. angew. Entomol.,* **24**, 367–435.

HÖLLDOBLER, K., 1938 b: *Zool. Anz.,* **121**, 66–72.

HÖLLDOBLER, K., 1944: *Z. angew. Entomol.,* **30**, 623–644.

HÖLLDOBLER, K., 1950: *Z. angew. Entomol.,* **31**, 583–590.

HÖLLDOBLER, K., 1953: *Z. angew. Entomol.,* **34**, 598–606.

HÖLLDOBLER, K., 1965: *Mitt. Schweiz. Entomol. Ges.,* **38**, 71–79.

HOMANN, H., 1924: *Z. vergl. Physiol.,* **1**, 541–578.

HORSTMANN, K., 1970: *Oecologia,* **5**, 138–157.

HORSTMANN, K., 1972: *Oecologia*, 8, 371–390.
HORSTMANN, K., 1973 a: *Z. Tierpsych.*, 32, 532–543.
HORSTMANN, K., 1973 b: *Waldhygiene*, 9, 193–201.
HORSTMANN, K., 1974 a: *Oecologia*, 15, 187–204.
HORSTMANN, K., 1974 b: *Waldhygiene*, 10, 241–246.
HORSTMANN, K., 1975 a: *Oecologia*, 22, 57–65.
HORSTMANN, K., 1975 b: *Waldhygiene*, 11, 33–40.
HORSTMANN, K., 1976: *Ins. soc.*, 23, 227–242.
HOWARD, D. F., W. T. TSCHINKEL, 1976: *Behav.*, 56, 157–180.
HUBER, J., 1905: *Biol. Cbl.*, 25, 606–619; 625–635.
HUBER, P., 1810: *Recherches sur les moeurs des fourmis indigènes.* Paris: J. J. Paschoud.
HUBER, P., 1814: *Nouvelles observations sur les abeilles.* Paris: J. J. Paschoud.
HUG, O., 1960: *Intern. J. Radiat. Biol. Suppl.*, 2, 217–226.
HUMMEL, H., P. KARLSON, 1968: *Hoppe Seyl. Z. physiol. Chem.*, 349, 725–727.
HUNG, A. C. F., H. T. IMAI, M. KUBOTA, 1972: *Ann. Entomol. Soc. Am.*, 65, 1023–1025.
HURD, P. D., R. F. SMITH, R. L. USINGER, 1956: *Proc. 10. Intern. Congr. Entomol., Montreal*, 1, 851.
HUWYLER, S., K. GROB, M. VISCONTINI, 1973: *Helv. chim. Acta*, 56, 976–977.
HUWYLER, S., K. GROB, M. VISCONTINI, 1975: *J. Ins. Physiol.*, 21, 299–304.

IHERING, H. v., 1894: *Berl. Entomol. Zschr.*, 39, 321–446.
IHERING, H. v., 1898: *Zool. Anz.*, 21, 238–245.

JACOBSON, E., A. FOREL, 1909: *Notes Leyden Mus.*, 31, 221–253.
JACOBY, M., 1953: *Z. angew. Entomol.*, 34, 145–169.
JACOBY, M., 1955: *Z. angew. Entomol.*, 37, 129–152.
JAISSON, P., 1969: *Ins. soc.*, 16, 279–312.
JAISSON, P., 1970: *C. R. Acad. Sc. Paris*, 271, 1192–1194.
JAISSON, P., 1972: *C. R. Acad. Sc. Paris*, 274, 302–305.
JAISSON, P., 1975: *Behav.*, 52, 1–37.
JANDER, R., 1957: *Z. vergl. Physiol.*, 40, 162–238.
JANDER, R., 1963: *Z. Tierpsych.*, 21, 302–307.
JANDER, R., 1970: *Z. Tierpsych.*, 27, 771–778.
JANET, C., 1898: *C. R. Acad. Sci. Paris*, 126, 1171–1186.

JANET, C., 1904: *Observations sur les fourmis.* Limoges: Ducourtieux et Gout.

JANZEN, D. H., 1966: *Evolution,* **20,** 249–275.

JANZEN, D. H., 1967 a: *Ecol.,* **48,** 26–35.

JANZEN, D. H., 1967 b: *Kans. Univ. Sci. Bull.,* **47,** 315–558.

JANZEN, D. H., 1969: *Ecol.,* **50,** 147–153.

JANZEN, D. H., 1972: *Ecol.,* **53,** 885–892.

JANZEN, D. H., 1973: *Ecol.,* **54,** 727–750.

JANZEN, D. H., 1975: *Science,* **188,** 936–937.

JOLIVET, P., 1973: *Cah. Pac.,* **17,** 41–69.

JOUVENAZ, D. P., W. A. BANKS, C. S. LOFGREN, 1974: *Ann. Entomol. Soc. Am.,* **67,** 442–444.

KANNOWSKI, P. B., R. L. JOHNSON, 1969: *Anim. Behav.,* **17,** 425–429.

KARAWAJEW, W., 1928: *Acad. Sc. l'Ucraine, Mém. Cl. Sci. Phys. Math.,* **6,** 307–328.

KARAWAJEW, W., 1929: *Zool. Anz.,* **82,** 247–256.

KARAWAJEW, W., 1931: *Zool. Anz.,* **92,** 309–317.

KARLSON, P., A. BUTENANDT, 1959: *Ann. Rev. Entomol.,* **4,** 39–58.

KARLSON, P., M. LÜSCHER, 1959: *Naturw.,* **46,** 63–64.

KAUDEWITZ, F., 1955: *Biol. Zbl.,* **74,** 69–87.

KELLY, A. F., 1973: *Ins. soc.,* **20,** 109–124.

KERMARREC, A., H. MAULÉON, A. ABUD ANTUN, 1976: *Ins. soc.,* **23,** 29–48.

KERR, W. E., 1962: *Arquivos Mus. Nation., Rio de Janeiro,* **52,** 115–116.

KIEPENHEUER, J., 1968: *Z. vergl. Physiol.* **57,** 409–411.

KING, R. L., F. WALTERS, 1950: *Proc. Iowa Acad. Sc.,* **57,** 469–473.

KISTNER, D. H., 1966: *Ann. Entomol. Soc. Am.,* **59,** 341–358.

KISTNER, D. H., 1967: *The Pan.-Pac.-Entomol.,* **43,** 274–284.

KISTNER, D. H., M. S. BLUM, 1971: *Ann. Entomol. Soc. Am.,* **64,** 589–594.

KLOFT, W., 1959 a: *Biol. Zbl.,* **78,** 863–870.

KLOFT, W., 1959 b: *Waldhygiene,* **3,** 94–98.

KLOFT, W., 1960: *Entomophaga,* **5,** 43–54.

KLOFT, W., B. HÖLLDOBLER, A. HAISCH, 1965: *Entomol. exp. et appl.,* **8,** 20–26.

KNEITZ, G., 1963: *Symp. Gen. Biol. Ital. Pavia,* **12,** 38–50.

KNEITZ, G., 1964: *Ins. soc.,* **11**, 105–130.

KNEITZ, G., 1967: *Progr. Soil Biol.,* 241–248.

KOHL, P. H., 1909: *Natur und Offenbarung,* **55**, 89–111; 148–175.

KÖHLER, F., 1966: *Ins. soc.,* **13**, 305–310.

KOOB, K., 1971: *Myrmecacin, das erste Insektenherbicid.* Diss. Univ. Heidelberg.

KRAUSSE, A. H., 1907: *Die antennalen Sinnesorgane der Ameisen, in ihrer Zahl und Verteilung bei den Geschlechtern und Individuen einiger Arten.* Diss. Univ. Jena.

KRAUSSE-HELDRUNGEN, A., 1910: *Zool. Anz.,* **35**, 523–526.

KUNKEL, H., 1972: *Bonn. Zool. Beitr.,* **23**, 161–178.

KUNKEL, H., 1973: *Bonn. Zool. Beitr.,* **26**, 105–121.

KÜRSCHNER, I., 1969: *Beitr. Entomol.,* **19**, 273–280.

KUTTER, H., 1921: *Mitt. Schweiz. Entomol. Ges.,* **13**, 117–120.

KUTTER, H., 1945: *Mitt. Schweiz. Entomol. Ges.,* **19**, 485–487.

KUTTER, H., 1948: *Mitt. Schweiz. Entomol. Ges.,* **21**, 286–295.

KUTTER, H., 1950: *Mitt. Schweiz. Entomol. Ges.,* **23**, 81–94.

KUTTER, H., 1951: *Mitt. Schweiz. Entomol. Ges.,* **24**, 153–174.

KUTTER, H., 1952: *Mitt. Schweiz. Entomol. Ges.,* **25**, 57–72.

KUTTER, H., 1956: *Mitt. Schweiz. Entomol. Ges.,* **29**, 1–18.

KUTTER, H., 1967: *Mitt. Schweiz. Entomol. Ges.,* **40**, 78–91.

KUTTER, H., 1969: *Neujahrsblatt, hrsg. von der Naturf. Ges. Zürich,* **171**, 1–62.

KUTTER, H., 1973 a: *Mitt. Schweiz. Entomol. Ges.,* **46**, 253–268.

KUTTER, H., 1973 b: *Mitt. Schweiz. Entomol. Ges.,* **46**, 281–289.

LAGERHEIM, G., 1900: *Entomol. Tidskr.,* **21**, 17–29.

LANGE, R., 1958: *Naturw.,* **45**, 196.

LANGE, R., 1959: *Entomophaga,* **4**, 47–55.

LANGE, R., 1960 a: *Z. angew. Entomol.,* **45**, 188–197.

LANGE, R., 1960 b: *Z. Tierpsych.,* **17**, 389–401.

LANGE, R., 1967: *Z. Tierpsych.,* **24**, 513–545.

LATREILLE, P., 1805: *Histoire naturelle générale et particulière des Crustacés et des Insectes, Band XIII.* Paris: Dufart.

LAW, J. H., E. O. WILSON, J. A. MAC CLOSKEY, 1965: *Science,* **149**, 544–546.

LEDOUX, A., 1949 a: *C. R. Acad. Sc. Paris,* **228**, 431–432.

LEDOUX, A., 1949 b: *C. R. Acad. Sc. Paris,* **228**, 1154–1155.

LEDOUX, A., 1950: *Ann. Sc. Nat. Zool.,* **12**, 313–461.

LEDOUX, A., 1952: *Ann. Sc. Nat. Zool.,* **14**, 231–248.

LEDOUX, A., 1954: *Ins. soc.,* **1,** 149–175.

LEDOUX, A., 1958: *Proc. 10. Int. Congr. Entomol. Montreal,* **2,** 521–528.

LEDOUX, A., 1971: *C. R. Acad. Sc. Paris,* **273,** 83–85.

LEDOUX, A., 1973: *C. R. Acad. Sc. Paris,* **277,** 2199–2200.

LEEUWEN, W. M. van, 1929: *Ber. Dtsch. Bot. Ges.,* **47,** 90–99.

LEHMANN, J., 1974: *Waldhygiene,* **10,** 252–255.

LEHMANN, J., 1975: *Waldhygiene,* **11,** 41–47.

LE MASNE, G., 1953: *Ann. Sc. Nat.,* **15,** 1–56.

LE MASNE, G., 1956 a: *Ins. soc.,* **3,** 239–259.

LE MASNE, G., 1956 b: *C. R. Acad. Sc. Paris,* **243,** 673–675.

LE MASNE, G., 1956 c: *C. R. Acad. Sc. Paris,* **243,** 1243–1246.

LE MASNE, G., A. BONAVITA, 1967: *Proc. 10. Int. Ethol. Conf. Stockholm.*

LE MASNE, G., C. TOROSSIAN, 1965: *Ins. soc.,* **12,** 185–194.

LEUTHOLD, R. H., 1968 a: *Psyche,* **75,** 233–248.

LEUTHOLD, R. H., 1968 b: *Psyche,* **75,** 334–350.

LEUTHOLD, R. H., 1975: *Symp. IUSSI, Dijon,* 197–211.

LEVIEUX, J., 1966: *Ins. soc.,* **13,** 117–126.

LEVIEUX, J., 1972: *Ins. soc.,* **19,** 63–79.

LINDAUER, M., H. MARTINI, 1963: *Naturw.,* **50,** 509–514.

LINDAUER, M., J. O. NEDEL, 1959: *Z. vergl. Physiol.,* **42,** 334–264.

LINSENMAIER, W., 1972: *Knaurs grosses Insektenbuch.* München– Zürich: Droemer Knaur.

LITTLEDYKE, M., J. M. CHERRETT, 1975: *Bull. Entomol. Res.,* **65,** 33–47.

LÖFQVIST, J., 1976: *J. Ins. Physiol.,* **22,** 1331–1346.

LOWE, G. H., 1948: *Proc. R. Entomol. Soc. London (A),* **23,** 51–53.

LUBBOCK, J., 1894: *Ants, bees and wasps.* New York: D. Appleton and Co.

LÜSCHER, M., 1956: *Ins. soc.,* **3,** 149–175.

MAC ALPINE, J. F., J. E. MARTINI, 1966: *Canad. Entomol.,* **98,** 527–544.

MAC COOK, H. C., 1877: *Trans. Am. Entomol. Soc.,* **6,** 253–296.

MAC COOK, H. C., 1880: *The natural history of the agricultural ant of Texas.* Philadelphia: Acad. of Natural Sciences.

MAIDL, F., 1934: *Die Lebensgewohnheiten und Instinkte der staatenbildenden Insekten.* Wien: Wagner.

MALICKY, H., 1969: *T. Entomol.,* **112,** 213–298.

MALYSHEV, S. I., 1966: *Genesis of the Hymenoptera and the phases of their evolution.* O. W. RICHARDS and B. UVANOV, eds. London: Methuen and Co.

MAMSCH, E., 1965: *Naturw.,* **52**, 168.

MAMSCH, E., 1967: *Z. vergl. Physiol.,* **55**, 1–25.

MAMSCH, E., K. BIER, 1966: *Ins. soc.,* **8**, 277–284.

MANN, W. M., 1919: *Bull. Mus. Comp. Zool.,* **63**, 273–391.

MARAK, G. E., J. J. WOLKEN, 1965: *Nature,* **205**, 1328–1329.

MARIKOVSKY, P. I., 1961: *Ins. soc.,* **8**, 23–30.

MARIKOVSKY, P. I., 1962: *Entomol. Rev.,* **41**, 47–51.

MARIKOVSKY, P. I., 1974: *Ins. soc.,* **21**, 301–308.

MARKIN, G. P., 1970: *Ann. Entomol. Soc. Am.,* **63**, 1238–1242.

MARKIN, G. P., H. L. COLLINS, J. H. DILLIER, 1972: *Ann. Entomol. Soc. Am.,* **65**, 1053–1058.

MARKL, H., 1962: *Z. vergl. Physiol.,* **45**, 475–569.

MARKL, H., 1964: *Z. vergl. Physiol.,* **48**, 552–586.

MARKL, H., 1965: *Science,* **149**, 1392–1393.

MARKL, H., 1966: *Verh. Dtsch. Zool. Ges., 30. Supp.,* 343–351.

MARKL, H., 1967: *Z. vergl. Physiol.,* **57**, 299–330.

MARKL, H., 1968: *Z. vergl. Physiol.,* **60**, 103–150.

MARKL, H., 1970: *Z. vergl. Physiol.,* **69**, 6–37.

MARKL, H., 1973 a: *Proc. VII. Congr. IUSSI, London,* 258–265.

MARKL, H., 1973 b: *Fortschr. Zool.,* **21**, 100–120.

MARKL, H., S. FUCHS, 1972: *Z. vergl. Physiol.,* **76**, 204–225.

MARLIN, J. C., 1968: *Trans. Ill. State Acad. Sc.,* **61**, 207–209.

MARLIN, J. C., 1969: *J. Kans. Entomol. Soc.,* **42**, 108–115.

MARLIN, J. C., 1971: *Am. Midl. Natural.,* **86**, 181–189.

MARTIN, M. M., 1969: *Science,* **169**, 16–20.

MARTIN, M. M., N. D. BOYD, M. J. GIESELMANN, R. G. SILVER, 1975: *Ins. Physiol.,* **21**, 1887–1892.

MARTIN, M. M., M. J. GIESELMANN, J. S. MARTIN, 1973: *J. Ins. Physiol.,* **19**, 1409–1416.

MARTIN, M. M., J. G. MAC CONNELL, G. R. GALE, 1969: *Ann. Entomol. Soc. Am.,* **62**, 386–388.

MARTIN, M. M., J. S. MARTIN, 1970: *J. Ins. Physiol.,* **16**, 109–119.

MARTIN, M. M., J. S. MARTIN, 1971: *J. Ins. Physiol.,* **17**, 1897–1906.

MARTIN, M. M., N. A. WEBER, 1969: *Ann. Entomol. Soc. Am.,* **62**, 1386–1387.

MARTINOYA, C., S. BLOCH, D. F. VENTURA, N. M. PUGLIA, 1975:

J. comp. Physiol., **104**, 205–210.

MARTINSEN, D. L., D. J. KIMELDORF, 1972: *Biol. Bull.,* **143**, 403–419.

MASCHWITZ, U., 1964 a: *Z. vergl. Physiol.,* **47**, 596–655.

MASCHWITZ, U., 1964 b: *Nature,* **204**, 324–327.

MASCHWITZ, U., 1971: *Naturw. Rundschau,* **24**, 485.

MASCHWITZ, U., 1974: *Oecologia,* **16**, 303–310.

MASCHWITZ, U., 1975 a: *Symp. IUSSI, Dijon,* 41–45.

MASCHWITZ, U., 1975 b: *Symp. IUSSI, Dijon,* 47–57.

MASCHWITZ, U., B. HÖLLDOBLER, 1970: *Z. vergl. Physiol.,* **66**, 176–189.

MASCHWITZ, U., B. HÖLLDOBLER, M. MÖGLICH, 1974: *Z. Tierpsych.,* **35**, 113–123.

MASCHWITZ, U., R. KLINGER, 1974: *Ins. soc.,* **21**, 163–166.

MASCHWITZ, U., W. KLOFT, 1971: In: W. BÜCHERL, E. BUCKLEY eds., *Venomous animals and their venoms,* 1–60. New York–London: Academic Press.

MASCHWITZ, U., E. MASCHWITZ, 1974: *Oecologia,* **14**, 289–294.

MASCHWITZ, U., M. MÜHLENBERG, 1973 a: *Zool. Anz.,* **191**, 364–368.

MASCHWITZ, U., M. MÜHLENBERG, 1973 b: Encycl. Cin., Göttingen.

MASCHWITZ, U., M. MÜHLENBERG, 1975: *Oecologia,* **20**, 65–83.

MASCHWITZ, U., M. WÜST, K. SCHURIAN, 1975: *Oecologia,* **18**, 17–21.

MASSON, C., M. A. FRIGGI, 1971 a: *C. R. Acad. Sc. Paris,* **272**, 618–621.

MASSON, C., M. A. FRIGGI, 1971 b: *C. R. Acad. Sc. Paris,* **272**, 2346–2349.

MASSON, C., D. GABOURIAUT, 1973: *Z. Zellf.,* **140**, 39–75.

MASSON, C., D. GABOURIAUT, A. FRIGGI, 1972: *Z. Morph. Tiere,* **72**, 349–360.

MAZOKHIN-PORSHNYAKOV, G. V., V. TRENN, 1972: *Zool. Zh.,* **51**, 1007–1017.

MENOZZI, C., 1924: *Soc. Nat. e Mat., Modena,* **6**, 6–8.

MENZEL, R., 1971: *Z. Naturf.,* **266**, 357.

MENZEL, R., 1972: *Z. Zellf.,* **127**, 356–373.

MENZEL, R., 1975: In: A. W. SNYDER, R. MENZEL, eds., *Photoreceptor optics,* 373–387. Berlin–Heidelberg–New York: Springer.

MENZEL, R., R. KNAUT, 1973: *J. comp. Physiol.*, **86**, 125–138.

MENZEL, R., R. WEHNER, 1970: *Z. vergl. Physiol.*, **68**, 446–449.

MENZER, G., K. STOCKHAMMER, 1951: *Naturw.*, **38**, 190.

MEYER, E., 1927: *Biol. Zbl.*, **47**, 264–307.

MEYER, G. F., 1955: *Ins. soc.*, **2**, 164–171.

MICHENER, C. D., D. J. BROTHERS, 1971: *Proc. Nat. Acad. Sc.*, **68**, 1241–1245.

MILNE, L. J., M. MILNE, 1976: *Scientific Am.*, **235**, 84–89.

MITTELSTAEDT-BURGER, M.-L., 1972: In: R. WEHNER, ed., *Information processing in the visual system of Arthropods.* Berlin–Heidelberg–New York: Springer.

MITTELSTAEDT, H., M.-L. MITTELSTAEDT, 1973: *Fortschr. Zool.*, **21**, 46–58.

MÓCZAR, L., 1972: *Acta Biol. Szeged*, **18**, 181–183.

MÖGLICH, M., 1971: *Nestumzugs- und Trageverhalten bei Ameisen.* Staatsexamensarbeit an der Univ. Frankfurt.

MÖGLICH, M., 1973: *Proc. VII. Congr. IUSSI, London*, 274–278.

MÖGLICH, M., B. HÖLLDOBLER, 1975: *J. comp. Physiol.*, **101**, 275–288.

MÖGLICH, M., U. MASCHWITZ, B. HÖLLDOBLER, 1974: *Science*, **186**, 1046–1074.

MÖLLER, A., 1893: *Bot. Mitt. Trop.*, 1–127.

MORGAN, E. D., L. J. WADHAMS, 1972: *J. Ins. Physiol.*, **18**, 1125–1135.

MORLEY, B. D. W., 1938: *Bull. Soc. Entomol. France*, **43**, 190–194.

MOSER, J. C., R. C. BROWNLEE, R. SILVERSTEIN, 1968: *J. Ins. Physiol.*, **14**, 529–535.

MÜHLENBERG, M., U. MASCHWITZ, 1973: Encycl. Cin., Göttingen.

MÜHLENBERG, M., U. MASCHWITZ, 1976: Encycl. Cin., Göttingen.

MÜLLER, E., 1931: *Z. vergl. Physiol.*, **14**, 348–384.

MÜLLER, F., 1874: *Nature*, **10**, 102–103.

MÜLLER, F., 1880: *Kosmos*, **8**, 109–116.

OESER, R., 1961: *Mitt. Zool. Mus. Berlin*, **37**, 1–119.

OFER, J., 1970: *Ins. soc.*, **17**, 49–82.

OLIVER, J. E., P. E. SONNET, 1974: *J. Org. Chem.*, **39**, 2662–2663.

OSMAN, M. F. H., W. KLOFT, 1961: *Ins. soc.*, **8**, 383–395.

OTTO, D., 1958: *Wiss. Abh. Dtsch. Akad. Landwirtschaftsw. Berlin*, **30**, 1–169.

OTTO, D., 1960: *Zool. Anz.*, **164**, 42–57.

OTTO, D., 1962: *Die Roten Waldameisen.* Wittenberg Lutherstadt: Ziemsen.

PARASCHIVESCU, D., 1967: *Ins. soc.,* **14**, 123–130.
PARK, O., 1933: *Ann. Entomol. Soc. Am.,* **26**, 255–261.
PASSERA, L., 1964: *Ins. soc.,* **11**, 59–70.
PASSERA, L., 1965: *C. R. 5. Congr. UIEIS, Toulouse,* 293–302.
PASSERA, L., 1966: *C. R. Acad. Sc. Paris,* **263**, 1600–1603.
PASSERA, L., 1969: *Ann. Sc. Nat. Zool.,* **11**, 327–481.
PASSERA, L., 1972: *C. R. Acad. Sc. Paris,* **275**, 409–411.
PASSERA, L., 1974: *Ins. soc.,* **21**, 71–86.
PASSERA, L., A. DEJEAN, 1974: *Ins. soc.,* **21**, 407–416.
PAULSEN, R., 1966: *Naturw.,* **53**, 337–338.
PAULSEN, R., 1969: *Zur Funktion der Propharynx-, Postpharynx- und Labialdrüsen von Formica polyctena Foerst. (Hymenoptera, Formicidae).* Diss. Univ. Würzburg.
PAULSEN, R., 1971: *Arch. Bioch. Biophys.,* **142**, 170–176.
PAVAN, M., 1956: *Ric. Sci.,* **26**, 144–150.
PEACOCK, A. D., J. H. SUDD, A. T. BAXTER, 1955: *Entomol. Month. Mag.,* **91**, 125–129.
PETERSEN, M., A. BUSCHINGER, 1971 a: *Anz. Schädlingsk.,* 121–127.
PETERSEN, M., A. BUSCHINGER, 1971 b: *Z. angew. Entomol.,* **68**, 168–175.
PETERSEN-BRAUN, M., A. BUSCHINGER, 1975: *Ins. soc.,* **22**, 51–66.
PIÉRON, H., 1906: *C. R. Acad. Sc. Paris,* **143**, 845–848.
PLSEK, R. W., J. C. KROLL, J. F. WATKINS, 1969: *J. Kans. Entomol. Soc.,* **42**, 452–456.
POLIMANTI, O., 1911: *Biol. Cbl.,* **31**, 222–224.
PONTIN, A. J., 1960 a: *Ins. soc.,* **7**, 227–230.
PONTIN, A. J., 1960 b: *Entomol. Month. Mag.,* **96**, 198–199.
PRELINGER, 1940: *Mikrokosm.,* **33**, 125–127.
PRIESNER, E., 1973: *Fortschr. Zool.,* **22**, 49–135.

QUILICO, A., F. PIOZZI, M. PAVAN, 1956: *Ric. Sci.,* **26**, 177–180.

RAIGNIER, A., 1959: *Med. Kon. VI. Acad. v. Belg.,* **21**, 24 S.
RAIGNIER, A., 1972: *Ins. soc.,* **19**, 153–170.
RAIGNIER, A., J. v. BOVEN, 1955: *Ann. Mus. Roy. Congo Belg.,*

N. S., Sc. Zool., **2**, 1–359.

REGNIER, F. E., M. NIEH, B. HÖLLDOBLER, 1973: *J. Ins. Physiol.*, **19**, 981–992.

REGNIER, F. E., E. O. WILSON, 1968: *J. Ins. Physiol.*, **14**, 955–970.

REGNIER, F. E., E. O. WILSON, 1971: *Science*, **172**, 267–269.

REICHENBACH, 1902: *Biol. Cbl.*, **22**, 461–465.

REICHENSPERGER, A., 1911: *Biol. Centralbl.*, **31**, 596–605.

RETTENMEYER, C.W., 1963: *Kans. Univ. Sc. Bull.*, **44**, 281–465.

REUTER, O. M., 1913: *Lebensgewohnheiten und Instinkte der Insekten bis zum Erwachen der sozialen Instinkte.* Berlin: Friedländer.

REYNE, A., 1954: *Zool. Mededel.*, **32**, 233–257.

RIBI, W. A., 1975: *Cell Tiss. Res.*, **160**, 207–217.

RIDLEY, H. N., 1890: *J. Str. Br. Roy. Asiat. Soc., Singapore*, 345

RILEY, R. G., R. M. SILVERSTEIN, B. CARROLL, R. CARROLL, 1974: *J. Ins. Physiol.*, **20**, 651–654.

RILEY, R. G., R. M. SILVERSTEIN, J. C. MOSER, 1974: *J. Ins., Physiol.*, **20**, 1629–1637.

RITTER, F. J., I. E. M. ROTGANS, E. TALMAN, P. E. J. VERWIEL, F. STEIN, 1973: *Experient.*, **29**, 530–531.

RITTER, F. J., I. E. M. BRÜGGEMANN-ROTGANS, E. VERKUIL, C. J. PERSOONS, 1975: *Symp. IUSSI, Dijon*, 99–103.

RITTER, F. J., F. STEIN, 1975: *Lokmiddel voor mieren.* Dutch patent application 74 08757 (28-6-1974).

ROBERTSON, P. L., 1968: *Austr. J. Zool.*, **16**, 133–166.

ROBERTSON, P. L., 1971: *J. Ins. Physiol.*, **17**, 691–715.

ROBINSON, S. W., J. M. CHERRETT, 1974: *Bull. Entomol. Res.*, **63**, 519–529.

ROBINSON, S. W., J. C. MOSER, M. S. BLUM, A. AMANTE, 1974: *Ins. soc.*, **21**, 87–94.

ROCKWOOD, L. L., 1976: *Ecology*, **57**, 48–61.

ROSENGREN, R., 1971: *Acta zool. fenn.*, **133**, 1–106.

ROTH, H., R. MENZEL, 1972: In: R. WEHNER, ed., *Information processing in the visual systems of Arthropods,* 177–181. Berlin–Heidelberg–New York: Springer.

RUTZ, W., L. GERIG, H. WILLE, M. LÜSCHER, 1976: *J. Ins. Physiol.*, **22**, 1485–1491.

SAMSINAK, K., 1964: *Acta Soc. Entomol. Cechoslov.*, **61**, 156–157.

SANDERS, C. J., 1972: *Proc. Entomol. Soc. Ontario,* **102,** 13–16.

SANTSCHI, F., 1907: *Rev. Suisse Zool.,* **15,** 305–334.

SANTSCHI, F., 1911: *Rev. Suisse Zool.,* **19,** 303–338.

SANTSCHI, F., 1920: *Rev. Zool. Afr.,* 7, 201–224.

SANTSCHI, F., 1923: *Mém. Soc. Vaudoise Sc. Nat.,* **1,** 137–176.

SAVILLE-KENT, W., 1897: *The naturalist in Australia.*

SCHENK, O., 1903: *Zool. Jhb.,* **17,** 573–618.

SCHILDKNECHT, H., K. KOOB, 1970: *Angew. Chem.,* **82,** 181.

SCHILDKNECHT, H., K. KOOB, 1971: *Angew. Chem.,* **83,** 110.

SCHIMPER, A. F. W., 1898: *Pflanzengeographie auf physiologischer Grundlage.* Jena: Fischer.

SCHMID, R., 1971: *Naturw. Rundsch.,* **24,** 482–485.

SCHMIDT, A., 1938: *Z. vergl. Physiol.,* **25,** 351–378.

SCHMIDT, G. H., 1972: *Zool. Anz.,* **189,** 159–169.

SCHMIDT, G. H., E. GÜRSCH, 1970: *Z. Morph. Tiere,* **67,** 172–182.

SCHMIDT, G. H., E. GÜRSCH, 1971: *Z. Tierpsych.,* **18,** 19–32

SCHMIDT, H., 1952: *Z. Morph. Ökol. Tiere,* **41,** 223–246.

SCHNEIDER, D., 1970: In: F. O. SCHMITT, Editor-in-Chief: *The Neurosciences: Second Study Program,* 511–518. New York: Rockefeller University Press.

SCHNEIDER, D., K.-E. KAISSLING, 1957: *Zool. Jhb. (Anat.),* **76,** 223–250.

SCHNEIDER, P., 1966: *Ins. soc.,* **13,** 297–304.

SCHNEIDER, P., 1971: *Zool. Anz.,* **187,** 202–213.

SCHNEIDER, P., 1972: *Ins. soc.,* **19,** 279–299.

SCHNEIRLA, T. C., 1944: *Am. Mus. Novit.,* **1253,** 1–26.

SCHNEIRLA, T. C., 1947: *Am. Mus. Novit.,* **1336,** 1–20.

SCHNEIRLA, T. C., 1948: *Zoologica,* **33,** 89–112.

SCHNEIRLA, T. C., 1952: *Cooloq. Int. Cent. Nat. Rech. Sc.,* **34,** 247–269.

SCHNEIRLA, T. C., 1971: *Army ants. A study in social organization,* H. R. TOPOFF, ed. San Francisco: W. H. Freeman and Comp.

SCHNEIRLA, T. C., A. Y. REYES, 1966: *Anim. Behav.,* **14,** 132–148.

SCHÖN, A., 1911: *Zool. Jhb. (Anatomie),* **31,** 439–472.

SCHUMACHER, A., W. G. WHITFORD, 1974: *Ins. soc.,* **21,** 317–330.

SEEVERS, C. H., 1965: *Fieldiana Zool.,* **47,** 137–351.

SMITH, M. R., 1939: *Proc. Entomol. Soc. Wash.,* **41,** 176–180.

SMITH, M. R., 1956: *Proc. Entomol. Soc. Wash.,* **58,** 271–275.

SNELLING, R. R., 1965: *Bull. S. Calif. Acad. Sc.,* **64,** 16–21.

SNYDER, A. W., R. MENZEL, 1975: *Photoreceptor optics.* Berlin—Heidelberg—New York: Springer.

SOMMER, E. W., R. WEHNER, 1975: *Cell Tiss. Res.,* **163**, 45—61.

SONNET, P. E., J. C. MOSER, 1972: *J. Agric. Food Chem.,* **20**, 1191—1194.

SONNET, P. E., J. C. MOSER, 1973: *Environm. Entomol.,* **2**, 851—854.

SOULIÉ. J., 1960 a: *Ins. soc.,* **7**, 283—295.

SOULIÉ, J., 1960 b: *Ins. soc.,* **7**, 369—376.

SOULIÉ, J., 1961: *Ins. soc.,* **8**, 213—297.

SOULIÉ, J., 1962: *Ins. soc.,* **9**, 181—195.

SOULIÉ, J., 1964: *Ins. soc.,* **11**, 383—388.

SPANGLER, H. G., 1967: *Science,* **155**, 1687—1689.

SPANGLER, H. G., 1974: *Ann. Entomol. Soc. Am.,* **67**, 458—460.

STÄGER, R., 1919: *Erlebnisse mit Insekten.* Zürich: Rascher und Co.

STÄGER, R., 1923: *Z. wiss. Insektenbiol.,* **18**, 290—292.

STÄGER, R., 1925: *Z. Morph. Ökol. Tiere,* **3**, 452—476.

STÄGER, R., 1929: *Zool. Anz.,* **82**, 177—184.

STAMMER, H.-J., 1938: *Z. angew. Entomol.,* **24**, 285—290.

STÄRCKE, A., 1936: *Entomol. Ber.,* **9**, 277—279.

STEINER, A., 1924: *Z. vergl. Physiol.,* **2**, 23—56.

STEINER, A., 1925:

STEINER, A., 1929: *Z. vergl. Physiol.,* **9**. 1—66.

STITZ, H., 1939: *Ameisen oder Formicidae.* In: F. DAHL, *Tierwelt Deutschlands.* Jena: Gustav Fischer.

STRICKLAND, A. H., 1951: *Bull. Entomol. Res.,* **42**, 65—103.

STUMPER, R., 1918: *Biol. Zbl.,* **38**, 160—179.

STUMPER, R., 1921 a: *Bull. Soc. Entomol. Belg.,* **3**, 24—30.

STUMPER, R., 1921 b: *Bull. Soc. Entomol. Belg.,* **3**, 90—97.

STUMPER, R., 1950: *Bull. Biol. France et Belg.,* **84**, 375—399.

STUMPER, R., 1953: *Bull. Soc. Nat. Luxemb.,* **46**, 130—135.

STUMPER, R., 1955 a: *Bull. Soc. Nat. Luxemb.,* **60**, 82—86.

STUMPER, R., 1955 b: *Bull. Soc. Nat. Luxemb.,* **60**, 87—97.

STUMPER, R., 1960: *Naturw.,* **47**, 457—463.

STUMPER, R., 1961: *Naturw.,* **48**, 735—736.

STUMPER, R., 1962: *Ins. soc.,* **9**, 329—333.

STUMPER, R., H. KUTTER, 1952: *C. R. Acad. Sc. Paris,* **234**, 1482—1485.

STURTEVANT, A. H., 1927: *Psyche,* **34**, 1—9.

SUDD, J. H., 1960: *Anim. Behav.*, **8**, 67—75.
SUDD, J. H., 1962: *Entomol. Month. Mag.*, **98**, 164—166.
SUDD, J. H., 1967 a: *Z. Tierpsych.*, **26**, 257—276.
SUDD, J. H., 1967 b: *An introduction to the behaviour of ants.*
  London: E. Arnold.
SUDD, J. H., 1970 a: *Ins. soc.*, **17**, 253—260.
SUDD, J. H., 1970 b: *Ins. soc.*, **17**, 261—272.
SWINTON, A. H., 1878: *Entomol. Month. Mag.*, **14**, 187.
SZLEP, R., T. JACOBI, 1967: *Ins. soc.*, **14**, 25—40.
SZLEP-FESSEL, R., 1970: *Ins. soc.*, **17**, 233—244.

TALBOT, M., 1959: *Am. Midl. Natural.*, **61**, 124—132.
TALBOT, M., 1971: *Psyche,* **78**, 169—179.
TALBOT, M., 1972: *J. Kans. Entomol. Soc.*, **45**, 254—258.
TANNER, J. E., 1892: *Trinid. Field Nat., Club,* **1**, 123—127.
TÄUBER, U., 1974: *J. comp. Physiol.*, **95**, 169—183.
TEPPER, J. G. O., 1882: *Trans. Proc. Roy. Soc. South Austr.*, **5**,
  24—26; 106—107.
TOHMEE, G., 1972: *Ins. soc.*, **19**, 95—103.
TOPOFF, H. R., 1971: *Am. Natural.*, **105**, 529—548.
TOPOFF, H. R., 1972: *Scientific Am.*, **227**, 3—11.
TORGERSON, R. L., R. D. AKRE, 1970 a: *Melandria (Wash. State
  Entomol. Soc.)* **5**, 1—28.
TORGERSON, R. L., R. D. AKRE, 1970 b: *J. Kans. Entomol. Soc.*,
  43, 395—404.
TOROSSIAN, C., 1959: *Ins. soc.*, **6**, 369—374.
TOROSSIAN, C., 1960: *Ins. soc.*, **8**. 383—391.
TOROSSIAN, C., 1961: *Ins. soc.*, **8**, 189—191.
TOROSSIAN, C., 1965: *C. R. Sanc. Soc. Biol.*, **159**, 984.
TOROSSIAN, C., 1966: *Ins. soc.*, **13**, 39—58.
TOROSSIAN, C., 1971 a: *Ins. soc.*, **18**, 193—202.
TOROSSIAN, C., 1971 b: *Ins. soc.*, **18**, 135—154.
TOROSSIAN, C., 1972: *Ins. soc.*, **19**, 25—38.
TRICOT, M. S., J. M. PASTEELS, B. TURSCH, 1972: *J. Ins. Physiol.*,
  18, 499—509.
TRIVERS, R. L., H. HARE, 1976: *Science,* **191**, 249—263.
TSCHINKEL, W. R., P. G. CLOSE, 1973: *J. Ins. Physiol.*, **19**,
  707—721.
TUMLINSON, J. H., R. M. SILVERSTEIN, J. M. MOSER, R. G.
  BROWNLEE, J. M. RUTH, 1971: *Nature,* **234**, 348—349.

TUMLINSON, J. H., R. M. SILVERSTEIN, J. M. MOSER, R. G.
  BROWNLEE, J. M. RUTH, 1972: *J. Ins. Physiol.*, 18, 809–814.

ULE, N., 1905: *Die Blumengärten der Ameisen am Amazonenstrom.*
  Vegetationsbilder, Jena, 3. Reihe.
ULE, N., 1906: *Ameisenpflanzen des Amazonasgebietes.*
  Vegetationsbilder, Jena, 4. Reihe.

VAN BOVEN, J. K. A., 1970: *Bull. Ann. Soc. R. Entomol. Belg.,* 106,
  127–132.
VANDERPLANK, F. L., 1960: *J. Anim. Ecol.,* 29, 15–33.
VIEHMEYER, H., 1912: *Entomol. Mitt.,* 1, 193–197.
VIEHMEYER, H., 1921: *Biol. Zbl.,* 41, 269–278.
VOSS, C., 1967: *Z. vergl. Physiol.,* 55, 225–254.
VOWLES, D. M., 1954 a: *J. exp. Biol.,* 31, 341–355.
VOWLES, D. M., 1954 b: *J. exp. Biol.,* 31, 356–375.
VOWLES, D. M., 1955: *Brit. J. Anim. Behav.,* 3, 1–13.

WALLIS, D. J., 1961: *Behav.,* 17, 17–47.
WALOFF, N., 1957: *Ins. soc.,* 4, 391–408.
WALSH, C. T., J. H. LAW, E. O. WILSON, 1965: *Nature,* 207,
  320–321.
WALSH, J. P., W. R. TSCHINKEL, 1974: *Anim. Behav.,* 22, 695–704.
WANG, Y. J., G. M. HAPP, 1974: *J. Ins. Morph. and Embryol.,* 3,
  73–86.
WARBURG, O., 1892: *Biol. Cbl.,* 12, 129–142.
WASMANN, E., 1889: *Wien. Entomol. Ztg.,* 8.
WASMANN, E., 1891: *Die zusammengesetzten Nester und gemischten*
  Kolonien der Ameisen. Münster/W.: Aschendorff.
WASMANN, E., 1893: *Biol. Cbl.,* 13, 39–40.
WASMANN, E., 1899: *Zoologica,* 26, 1–133.
WASMANN, E., 1901: *Allgem. Entomol.,* 2, 19.
WASMANN, E., 1905: *Notes Leyden Mus.,* 25, 133–140.
WASMANN, E., 1909: *Biol. Cbl.,* 29, 587–604; 619–637; 651–663;
  683–703.
WASMANN, E., 1910 a: *Biol. Cbl.,* 30, 453–464; 475–496; 515–524.
WASMANN, E., 1910 b: *Dtsch. Entomol. Nationalbibl.,* 1, 1–11.
WASMANN, E., 1913: *Biol. Cbl.,* 33, 264–266.
WASMANN, E., 1915: *Z. wiss. Zool.,* 144, 233–402.

WASMANN, E., 1920: *Die Gastpflege der Ameisen.* Berlin: Gebr. Bornträger.

WASMANN, E., 1934: *Die Ameisen, die Termiten und ihre Gäste.* Regensburg: G. J. Manz AG.

WATKINS, J. F., T. W. COLE, 1966: *Texas J. Sc.,* **18,** 254−265.

WATKINS, J. F., F. R. GEHLBACH, R. S. BALDRIDGE, 1967: *Southw. Nat.,* **12,** 455−462.

WATKINS, J. F., F. R. GEHLBACH, J. C. KROLL, 1969: *Ecology,* **50,** 1099−1102.

WAY, M. J., 1954 a: *Bull. Entomol. Res.,* **45,** 93−112.

WAY, M. J., 1954 b: *Bull. Entomol. Res.,* **45,** 113−134.

WAY, M. J., 1963: *Ann. Rev. Entomol.,* **8,** 307−344.

WEATHERSTON, J., 1967: *Quart. Rev.,* **21,** 281−313.

WEBER, N. A., 1943: *Ecology,* **24,** 400−404.

WEBER, N. A., 1947: *Bol. Entomol. Venez.,* **6,** 143−161.

WEBER, N. A., 1949: *Ecology,* **30,** 397−400.

WEBER, N. A., 1956 a: *Ecology,* **37,** 150−161.

WEBER, N. A., 1956 b: *Anat. Rec.,* **125,** 604−605.

WEBER, N. A., 1958: *Proc. 10. Int. Congr. Entomol., Montreal,* **2,** 459−473.

WEBER, N. A., 1966: *Science,* **153,** 587−604.

WEBER, N. A., 1972 a: *Mem. Am. Phil. Soc.,* **92,** 1−146.

WEBER, N. A., 1972 b: *Am. Zoologist,* **12,** 577−587.

WEHNER, R., 1968: *Rev. Suisse Zool.,* **75,** 1076−1085.

WEHNER, R., 1970: *Ins. soc.,* **17,** 83−93.

WEHNER, R., 1975 a: *Fortschr. Zool.,* **23,** 148−160.

WEHNER, R., 1975 b: *Umschau Wiss. Techn.,* **75,** 653−660.

WEHNER, R., P. L. HERRLING, A. BRUNNERT, R. KLEIN, 1972: *Rev. Suisse Zool.,* **79,** 197−228.

WEHNER, R., P. LUTZ, 1969: *Natur und Museum,* **99,** 177−190.

WEHNER, R., F. TOGGWEILER, 1972: *J. comp. Physiol.,* **77,** 239−255.

WEILER, P., 1936: *Fühleruntersuchengen an Ameisen, insbesondere an Dorylinae.* Diss. Univ. Bonn.

WEIR, J. S., 1958 a: *Ins. soc.,* **5,** 97−128.

WEIR, J. S., 1958 b: *Ins. soc.,* **5,** 315−339.

WEIR, J. S., 1959: *Ins. soc.,* **6,** 271−290.

WELLENSTEIN, G., 1952: *Z. Pflanzenkr.,* **52,** 430−451.

WELLENSTEIN, G., 1957: *Z. angew. Entomol.,* **41,** 368−385.

WERRINGLOER, A., 1932: *Z. wiss. Zool.,* **141,** 432−524.

WESSON, L. G., 1939: *Trans. Am. Entomol. Soc.*, **65**, 97–122.
WESSON, L. G., 1940: *Bull. Brookl. Entomol. Soc.*, **35**, 73–83.
WEYER, F., 1927: *Zool. Anz.*, **74**, 205–221.
WEYER, F., 1928: *Z. wiss. Zool.*, **131**, 345–501.
WEYER, F., 1936: *Z. Morph. Ökol. Tiere*, **30**, 629–634.
WHEELER, G. C., J. WHEELER, 1970: *Ann. Entomol. Soc. Am.*, **63**, 648–656.
WHEELER, W. M., 1901: *Am. Nat.*, **35**, 513–539.
WHEELER, W. M., 1903: *J. Psychol. Neurol.*, **2**, 1–31.
WHEELER, W. M., 1904 a: *Bull. Am. Mus. Nat. Hist.*, **20**, 347–375.
WHEELER, W. M., 1904 b: *Biol. Bull.*, **6**, 251–259.
WHEELER, W. M., 1907: *Bull. Am. Mus. Nat. Hist.*, **23**, 669–807.
WHEELER, W. M., 1910: *Ants: Their structure, development and behaviour.* Columbia Univ. Press.
WHEELER, W. M., 1914: *Schr. Phys.-ökon. Ges. Königsberg*, **55**, 1–142.
WHEELER, W. M., 1915: *Ann. Entomol. Soc. Am.*, **8**, 323–342.
WHEELER, W. M., 1919: *Proc. Am. Phil. Soc.*, **58**, 1–40.
WHEELER, W. M., 1921: *Ecology*, **2**, 89–103.
WHEELER, W. M., 1923: *Social life among the insects.* London–Bombay–Sydney: Constable and Comp. Ltd.
WHEELER, W. M., 1926: *Quart. Rev. Biol.*, **2**, 1–36.
WHEELER, W. M., 1928: *The social insects: Their origin and evolution.* London: Kegan Paul, Trench, Trubner and Co., Ltd.
WHEELER, W. M., 1932: *Science*, **76**, 532–533.
WHEELER, W. M., 1933: *Colony-founding among ants.* Cambridge/Mass.: Harv. Univ. Press.
WHEELER, W. M., 1936: *Proc. Am. Acad. Arts Sc.*, **71**, 159–243.
WHEELER, W. M., J. C. BEQUAERT, 1929: *Zool. Anz.*, **82**, 10–39.
WHELDEN, R. M., 1957: *Ann. Entomol. Soc. Am.*, **50**, 271–282.
WHELDEN, R. M., 1960: *Ann. Entomol. Soc. Am.*, **53**, 793–808.
WHITCOMB, W. M., A. BHATKAR, J. C. NIKKERSON, 1973: *Environm. Entomol.*, **2**, 1101–1103.
WHITFORD, W. G., P. JOHNSON, J. RAMIREZ, 1976: *Ins. soc.*, **23**, 117–132.
WICKLER, W., 1968: *Mimikry. Nachahmung und Täuschung in der Natur.* München: Kindlers Univ. Bibl.
WILSON, E. O., 1954: *Ins. soc.*, **1**, 75–80.
WILSON, E. O., 1955: *Bull. Comp. Zool.*, **113**, 1–201.

WILSON, E. O., 1958 a: *Evolution,* **12,** 24–31.
WILSON, E. O., 1958 b: *Psyche,* **65,** 41–51.
WILSON, E. O., 1958 c: *Ins. soc.,* **5,** 129–140.
WILSON, E. O., 1959: *Psyche,* **66,** 29–34.
WILSON, E. O., 1962 a: *Bull. Mus. Comp. Zool.,* **127,** 403–422.
WILSON, E. O., 1962 b: *Anim. Behav.,* **10,** 134–147.
WILSON, E. O., 1962 c: *Anim. Behav.,* **10,** 148–158.
WILSON, E. O., 1963: *Ann. Rev. Entomol.,* **8,** 345–368.
WILSON, E. O., 1965: *Psyche,* **72,** 2–7.
WILSON, E. O., 1971: *The insect societies.* Cambridge/Mass.: The
  Belknap Press of Harv. Univ. Press.
WILSON, E. O., 1975 a: *Insects, Science, and Society,* 25–31.
WILSON, E. O., 1975 b: *Sociobiology, the new synthesis.*
  Cambridge/Mass.: The Belknap Press of Harv. Univ. Press.
WILSON, E. O., 1976: *Anim. Behav.,* **24,** 354–363.
WILSON, E. O., W. H. BOSSERT, 1963: *Rec. Prog. Horm. Res.,* **19,**
  673–716.
WILSON, E. O., W. L. BROWN, 1956: *Ins. soc.,* **3,** 439–454.
WILSON, E. O., F. M. CARPENTER, W. L. BROWN, 1967: *Science,*
  157, 1038–1040.
WILSON, E. O., N. DURLACH, L. M. ROTH, 1958: *Psyche,* **65,**
  108–114.
WILSON, E. O., T. EISNER, 1957: *Ins. soc.,* **4,** 157–166.
WILSON, E. O., T. EISNER, B. D. VALENTINE, 1954: *Psyche,* **61,**
  154–160.
WILSON, E. O., T. EISNER, G. C. WHEELER, J. WHEELER, 1956:
  Bull. Mus. Comp. Zool., **115,** 81–98.
WILSON, E. O., M. PAVAN, 1959: *Psyche,* **66,** 70–76.
WILSON, E. O., R. W. TAYLOR, 1964: *Psyche,* **71,** 93–103.
WRAY, D. L., 1938: *Ann. Entomol. Soc. Am.,* **31,** 196–201.
WROUGHTON, R. C., 1892: *J. Bombay Nat. Hist. Soc.,* 13–60.

YARROW. I. H. H., 1955: *Proc. Roy. Entomol. Soc. London,* **24,**
  113–115.
YARROW, I. H. H., 1968: *The Entomolog.,* 236–240.

ZAHN, M., 1958: *Zool. Beitr. N. F.,* **3,** 127–194.
ZEBITZ, C., 1979: *Ins. soc.,* **26,** 1–4.

ZOLOTOV, V., L. FRANTSEVICH, 1973: *J. comp. Physiol.,* **85,**
25–36.
ZWÖLFER, H., 1958: *Z. angew. Entomol.,* **43,** 1–52.

Index